Synthesis Lectures on Mechanical Engineering

This series publishes short books in mechanical engineering (ME), the engineering branch that combines engineering, physics and mathematics principles with materials science to design, analyze, manufacture, and maintain mechanical systems. It involves the production and usage of heat and mechanical power for the design, production and operation of machines and tools. This series publishes within all areas of ME and follows the ASME technical division categories.

Michał Stosiak · Mykola Karpenko

Dynamics of Machines and Hydraulic Systems

Mechanical Vibrations and Pressure Pulsations

 Springer

Michał Stosiak [ID]
Department of Technical Systems Operation
and Maintenance
Faculty of Mechanical Engineering
Wrocław University of Science
and Technology
Wrocław, Poland

Mykola Karpenko [ID]
Department of Mobile Machinery and Railway
Transport
Faculty of Transport Engineering
Vilnius Gediminas Technical University
Vilnius, Lithuania

ISSN 2573-3168   ISSN 2573-3176   (electronic)
Synthesis Lectures on Mechanical Engineering
ISBN 978-3-031-55524-4   ISBN 978-3-031-55525-1   (eBook)
https://doi.org/10.1007/978-3-031-55525-1

This Springer imprint is published by the registered company Springer Nature Switzerland AG
The registered company address is: Gewerbestrasse 11, 6330 Cham, Switzerland

Paper in this product is recyclable.

Preface

The subject of this book is to examine the influence of mechanical vibration on the changes in the pressure pulsation spectrum of hydraulic systems. In book displayed that machines and equipment equipped with hydraulic systems are a source of vibration with a wide frequency spectrum. Additionally, hydraulic valves are also exposed to vibration. Vibrations of the substrate on which the hydraulic valve is installed force the control element of the hydraulic valve to vibrate. The control element's vibration produced in this way causes changes in the pressure pulsation spectrum of the hydraulic system. A friction model modified using mixed friction theory can be used for the oscillating motion of the hydraulic directional control spool. Passive vibration isolation methods are proposed to reduce valve vibration. The biomimetic approach can be implemented in hydraulic systems (for pipeline) to reduce mechanical vibration and fluid pulsation. Numerical methods are employed to analyze the effect of changes in the pressure pulsation spectrum on the hydraulic efficiency of the pipelines. Examples are provided for the implementation of numerical methods in the calculation of hydraulic components and systems. Additionally, the effects of energy-saving in hydraulic systems by applying proposed results overview in the current book. The current book will be interesting for both—scientifical and manufacturing stuff—since the implementation of knowledge can help to design more substantial construction of machine hydraulic systems with avoiding vibration problems.

Chapter 1 is a concise classification of the term vibration, as well as its sources and selected effects. We emphasize that mechanical vibration that occurs in the working environment of the machine adversely affects the hydraulic valves, the machine, and its human operator. The frequency spectrum of the acting vibrations has been highlighted, with a particular focus on low frequencies, infrasonic frequencies, and frequencies close to the natural frequency of the hydraulic valve controls. Based on reported experiments, Chap. 1 also addresses the excitation of mechanical vibration of hydraulic lines caused by the pulsating flow of the working fluid. To demonstrate the coincidence between pressure pulsation in the hydraulic system and mechanical vibrations of the system's components, a constructed hydraulic pulsator stand is presented based on the latest generation proportional valve. The experimental results prove the occurrence of vibration of the body of the tested proportional directional valve. The vibration is induced by the pulsation of pressure in the hydraulic system, where reports focus on the frequency range <100 Hz. This problem is referred to in the literature as flow-induced vibration (FIV). Additionally, Chap. 1 explores energy-saving strategies in hydraulic systems through the mitigation of dynamic loads, focusing on mechanical and hydraulic vibrations. It delves into the realm of mechanical vibrations in machinery equipped with hydraulic systems, offering insights into their classification.

Chapter 2 identifies the sources of pressure pulsations in hydraulic systems. Four basic sources are specified: pulsation of displacement pumps efficiency, transient states, the influence of external mechanical vibrations on hydraulic system elements, and the wave phenomena in hydraulic lines of a long hydraulic line (hydraulic transmission line). Based on literature analysis, the instantaneous efficiency of the basic types of positive displacement pumps (single-acting vane and external tooth meshing) has been determined using analytical relationships. Computer simulation models have been created using the available modules of computer software. To identify the frequency spectrum of the capacity pulsation, the time waveforms and the amplitude-frequency spectra are presented. This chapter shows that an inseparable stage of the hydraulic system's operation is its start-up or braking (sometimes temporary), which is classified by transient processes. Furthermore, unlike pressure pulsation caused by the pulsation of the capacity of positive displacement pumps, the spectrum of pressure pulsations during the system start-up also includes low-frequency harmonics. We provide an example of the coincidence of external mechanical vibration and pressure pulsations for a micro-relief valve and a conventional one-stage relief valve. The occurrence of resonance phenomena in the hydraulic lines at quasi-steady flows is also described: the conditions for the occurrence of such phenomena and their effects are given. The quasi-steady friction model is combined with the concept of a hydraulic cross involving four items, Laplace transforms of pressure p_1, flow rate q_1 at the beginning of the hydraulic transmission line, and Laplace transforms of pressure and flow rate at the end of the hydraulic transmission lines p_2 and q_2, respectively. These tools are utilized to determine the transmittance maxima in the hydraulic system, and for the selected and parameterized hydraulic system, the resonance lengths of the lines are determined that increase the amplitude of pressure pulsation due to quasi-steady flow.

Chapter 3 is dedicated to employing numerical techniques for the assessment of the influence of fluid flow pulsations on vibrations in hydraulic lines. Throughout this chapter, we explore the mathematical representation of heightened pulsations in hydraulic system lines, employing the axial-piston pump model. Additionally, we elaborate on the mathematical depiction of high-pressure hose behavior, contingent upon fluid flow. This chapter introduces a fluid-solid coupling mathematical framework, integrating mechanical equations resolved through the finite elements method (FEM) and hydrodynamic equations addressed via the method of characteristics (MoC). These two sets of equations are concurrently solved, resulting in a comprehensive model designed to enhance our understanding of the complexities associated with fluid pulsations within hydraulic drive lines.

Chapter 4 examines the transmission pathways of external mechanical vibrations to the control element of the hydraulic valve (e.g., lift valve poppet or directional spool). For this purpose, we present common mathematical models of friction in motion pairs occurring in hydraulic elements. The experimental results of work conducted on the value related to friction force in the spool pair are shown. Then, based on our own theoretical and experimental considerations, we attempt to identify a significant damping parameter in the transmission of mechanical vibration to the valve control element. Our results demonstrate the excitation of vibrations of the valve control element by the external mechanical vibration and the related changes in the amplitude-frequency spectrum of pressure pulsations of the system. The influence of the direction of mechanical vibration on the excitation of the control element of the hydraulic directional valve is presented. Correlation is shown between the external mechanical vibrations of the valve and changes in the amplitude-frequency spectrum of pressure pulsations in the hydraulic system with the valve exposed to vibration, depending on the direction of external vibration. This problem is referred to as fluid-structure interaction (FSI) in the international nomenclature and covers more complex issues than the previously mentioned FIV. Mathematical models of cooperation of the hydraulic spool pair of the hydraulic proportional valve are refined using elements of mixed friction theory. A mathematical description closest to the experimental data has been determined based on the experimental verification studies. The parameters of the mathematical models are estimated using special computer tools.

Chapter 5 focuses on the possibilities of reducing the impact of mechanical vibration on hydraulic valves: on their body and the control element. This is examined theoretically and experimentally. Passive methods of isolating the valve body from vibration are considered using a specially designed and manufactured holder that simulates the machine frame, in which the hydraulic valve is mounted in a flexible manner. Flexible mounting of the valve consists of inserting a spring pack or a set of elastomer washers (or made of oil-resistant rubber) between the ground (special holder) and the valve body, on one or both sides of the valve's body. The elastomer washers are introduced by placing the washers between a special handle and the valve body, or specially profiled washers are inserted inside the valve, between the springs centering the spool and the body. Conclusions from the experimental testing were extended by the theoretical analysis of the possibilities of vibration reduction carried out in the further part of Chap. 7. Additionally, the present chapter unveils the research findings on a suggested energy-saving method in hydraulic drives rooted in a biomimetic approach. The study involves experimental measurements aimed at investigating energy consumption across various types of high-pressure hoses (pipelines) and assessing its impact on internal fluid flow dynamics. In the chapter, approximate analytical methods for the analysis of non-linear vibration reduction systems are provided, which can be applied to hydraulic valves.

Chapter 6 presents the research results of the proposed energy-saving way in hydraulic drives based on a reducing vibration. Additionally, the effects of energy-saving in hydraulic systems by applying proposed results overview in the current chapter. The current book chapter held present how implementation of knowledge can help to design more substantiable construction of machine hydraulic systems with avoiding vibration problems.

Chapter 7 shows the general observations on transmission paths, as well as the impact and the possibility of reducing the impact of external mechanical vibration on hydraulic valves. A simplified, practical scheme for selecting the hydraulic valve's vibration isolation is also provided.

Appendix provides extended results of testing the effects of external mechanical vibrations on a single-stage pressure relief valve and a single-stage conventional 4/3 spool valve. The results are presented in the form of amplitude-frequency spectra of pressure pulsations in the hydraulic system, in which the tested valve vibrates harmonically with a specific frequency and amplitude.

Wrocław, Poland   Michał Stosiak
Vilnius, Lithuania   Mykola Karpenko

Contents

About the Authors

Michał Stosiak was born on August 10, 1977 in Wrocław (Poland). In 2001, at Wrocław University of Science and Technology, he acquired Master's degree in Mechanical Engineering. In 2005 defended Ph.D. dissertation at Wrocław University of Science and Technology and obtained Doctoral degree in Mechanical Engineering, construction and operation of machinery, hydraulic drive and control. In 2015 defended Habilitation at Wrocław University of Science and Technology and obtained Doctor of Science (Dr. Habil.) degree in Mechanical Engineering. From 2017 working in Associate Professor position at the Mechanical Engineering Faculty of Wrocław University of Science and Technology. Deputy Head of Department of Technical Systems Operation and Maintenance and former Head of Department of Hydraulic Drives and Automatics. Author of more than 300 scientific publications in Mechanical Engineering field and author of seven patents and six patents applications. The main research direction interests include: hydraulics, vibrations, pressure pulsation and damping, noise, hydrodynamics, frequency and vibration analysis, and application of composite materials.

Mykola Karpenko was born on May 29, 1994 in Myronovka (Kyiv region, Ukraine). In 2015, at Kyiv National University of Construction and Architecture, he acquired Master's degree in Mechanical Engineering. In 2021 defended Ph.D. dissertation at Vilnius Gediminas Technical University and obtained Doctoral degree in Transport Engineering. From 2022 working in Associate Professor position at the Transport Engineering Faculty of Vilnius Gediminas Technical University. Author of more than 60 scientific publications in Mechanical and Transport Engineering fields. The main research direction interests include: hydrodynamics, numerical simulation, FEA and CFD, and frequency and vibration analysis.

Vibrations in Machines Fitted with Hydraulic Systems

1.1 Introduction

The utilization of fluid power in hydraulic systems has witnessed a transformative evolution over the past century. In 1906, oil emerged as a substitute for water as the pressurized fluid in hydraulic systems, opening up a new possibilities for industrial applications. The introduction of servo valves during World War II further revolutionized the field by enabling electrical control of fluid power, paving the way for its integration into the commercial sector. As a result, the fluid power industry has grown into a thriving multibillion-dollar sector today. In the extraction industries, such as mining, road construction, logging, farming, and transportation, fluid power circuits have become an integral part of most mobile machines. This widespread adoption has rendered fluid power a crucial element in the global economy's raw materials collection process. Moreover, hydraulic drives have proven indispensable in the transportation industry due to their ability to generate substantial force and torque within systems.

However, hydraulic systems encounter challenges during operation, particularly in dealing with various types of vibrations. These vibrations range from fluid flow-induced vibrations to internal and external machine-related vibrations. Additionally, hydraulic drives employed in construction and mobile machines suffer from high-energy consumption and low efficiency, typically achieving only around 60–70% efficiency. To address these issues, the European Union has adopted strategies like "Transport 2050" [1] to improve energy efficiency and sustainability. As a response to the growing importance of energy-saving measures in European strategies, extensive research is currently directed towards increasing the efficiency of hydraulic drives. One promising approach involves mitigating different types of vibrations in hydraulic systems to enhance overall performance. Consequently, in-depth research into the types and potential methods of vibration damping in hydraulic drives has become a prominent and contemporary in pursuit of

© The Author(s), under exclusive license to Springer Nature Switzerland AG 2024
M. Stosiak and M. Karpenko, *Dynamics of Machines and Hydraulic Systems*, Synthesis Lectures on Mechanical Engineering, https://doi.org/10.1007/978-3-031-55525-1_1

energy-saving goals. By investigating these aspects, researchers aim to make substantial contributions to enhancing the efficiency and reliability of hydraulic drives, aligning with the objectives set forth by energy-conscious European initiatives.

This book chapter explores energy-saving strategies in hydraulic systems through the mitigation of dynamic loads, focusing on mechanical and hydraulic vibrations. It delves into the realm of mechanical vibrations in machinery equipped with hydraulic systems, offering insights into their classification. The chapter also examines how machines can serve as both generators and recipients of mechanical vibrations, emphasizing their dual role. Furthermore, the text delves into fluid pulsations (vibrations) within hydraulic systems of machines. It discusses the impact of fluid pulsation on pipelines and highlights the consequential influence on hydraulic system performance.

1.1.1 Energy-Saving in Hydraulic Systems by Reduce Dynamic Loads—Mechanical and Hydraulic Vibrations

Fluid-based power technology involves the process of converting mechanical energy into liquid energy, transmitting this energy to a designated point of use, and subsequently converting it back into mechanical energy. A fluid power circuit incorporates all three essential elements: the conversion of mechanical energy to liquid energy, the delivery of this fluid energy, and the reconversion of the fluid energy back into mechanical energy. This process occurs simultaneously, even when subject to varying types of vibration (refer to Fig. 1.1).

Various construction, building, transport, and road machines have been developed using diverse design concepts and approaches. According to Helbig [2], these machines utilize a hydraulic drive system for powering their working equipment. This hydraulic drive is achieved through power transmission, which can consume a significant portion of the energy generated by the primary combustion engine or electric motor during the machine's working cycle. The energy consumption for the power transmission can range from 30% to as high as 90% of the total energy output from the combustion engine or electric motor. The hydraulic drive remains relevant in three main areas of energy conservation, which can be further divided into several sub-directions: Regulation and control of hydraulic drives; Increasing the efficiency of hydraulic drives through fluid recovery; Reducing

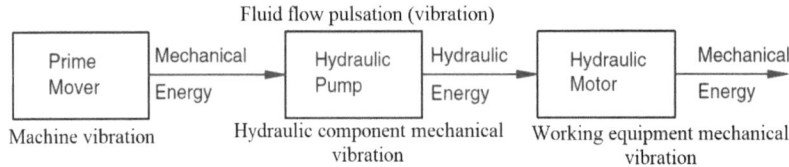

Fig. 1.1 Concept of hydraulic drive fluid power circulation and vibration types

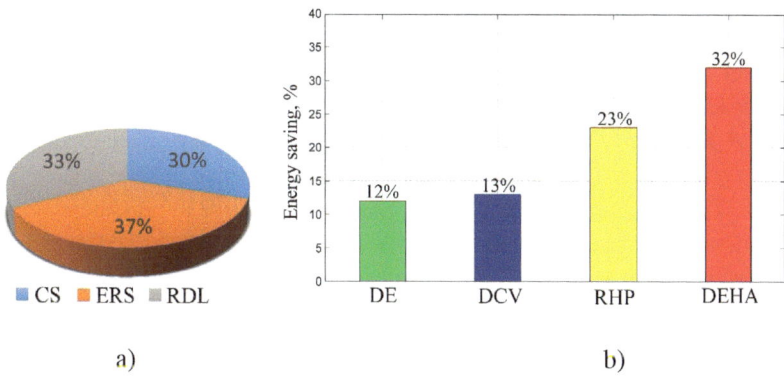

Fig. 1.2 Energy study in hydraulic systems: **a** energy-saving methods compering; **b** reducing dynamic loads (vibration)

dynamic loads in the system. Karpenko et al. [3] conducted a detailed review of energy-saving research in hydraulic drives, revealing the following popular research directions and their respective percentages:

- Usage of energy recovery systems (ERS) (37%);
- Methods for reducing dynamic loads in a system (RDL) (33%);
- Control of systems (CS) (30%).

Each of these reviewed methods is nearly equivalent in importance, as depicted in Fig. 1.2a. According to Borghia et al. [4] only RDL method can be implemented in nearly any hydraulic drive with minimal modifications.

In the research works by Adachi et al. [5] and Wang, Huang [6] have suggested that an effective approach to lower power consumption in hydraulic drives is by minimizing dynamic effects within the system. Stosiak [7] also highlights the adverse consequences of periodic pressure fluctuations in hydraulic systems, such as excessive noise emission, reduced component lifespan, disruptions in control loops, power loss (pressure) within the system, and more. To address this type of problem, one common approach involves incorporating elastic-damping elements into the hydraulic drive, as proposed by Ortwig [8]. Additionally, hydraulic accumulators and various fluid flow dampers, as indicated in the works of Han and Wang [9], and Amirante et al. [10], can be utilized to reduce dynamic loads in hydraulic drives.

For instance, Makaryants et al. [11] focused on mitigating fluid pressure pulsations and hydraulic resistance in a punching machine equipped with a high-pressure hose near the pump. By installing a damping element, they achieved a remarkable reduction of hydraulic pulsations in the system, resulting in about 38...40% decrease from the initial levels. This reduction in pulsations led to a decrease in pressure loss and contributed to approximately 13% energy savings in the hydraulic drive system. In test benches with harmonic vibrations, proposed by Yakushev [12], the feed pump's supply can be reduced by 30% with reducing fluid fluctuations by installing in the pressure line air-hydraulic accumulator for saving energy more than 23%.

Certainly! The utilization of a hydraulic accumulator in power cylinders of a manipulator, with a payload capacity of 0.8 t, has proven to be effective in reducing energy costs by 7…12%. This reduction is attributed to the alleviation of peak pressure spikes within the hydraulic system during transient regimes, resulting in a diminished dynamic effect on the hydraulic pump. This research was conducted by Nesmiyanov and Khavronin in 2007 [13]. Furthermore, when short-term high-speed mechanisms (ejectors) are present in the hydraulic system, installing damping elements directly near the cylinders has shown promising energy-saving results. This installation approach eliminates vibrations in long high-pressure hoses, contributing to enhanced energy efficiency within the hydraulic system.

According to Shen et al. [14], one of the drawbacks of active vibration compensation systems is that the energy introduced into the hydraulic system through the actuator can negatively impact the overall system stability. In cases where the regulator is not precisely tuned, there is a risk of deteriorating the system's characteristics, and under certain conditions, the amplitude of pressure fluctuations may actually increase instead of being reduced. Hence, careful and accurate tuning of the regulator is crucial to ensure the effectiveness and stability of active vibration compensation systems in hydraulic applications.

The comperes of using different types of reducing dynamic loads in the system for energy-saving shown in Fig. 1.2b, where: DE—Damping element; DCV—Damping of the cylinder and valves; RHP—Reduce hydraulic pulsation; DEHA—Damping effect by use hydraulic accumulator.

From another point of view, pressure fluctuations can occur either in the hydraulic systems themselves (fluid vibration) or due to external causes (mechanical vibration), for example, due to periodic fluctuations in load on hydraulic cylinders, valves, pipeline etc. [15]. Also known that in hydraulic systems with high dynamics (for example, proportional valve or servo valve) and with hydraulic cylinder or motor, there maybe a too high-pressure level of fluctuation what leads to energy losses mentioned by Stosiak [7]. In this case the detail research on the mechanical and fluid vibration characteristics in hydraulic systems and possible of its reducing currently one of the commom problem for solving in line obtain energy-saving by minimal modifications of hydraulic system.

1.1.2 Mechanical Vibrations in Machines Fitted with Hydraulic Systems

In this section, the main focus is provided on free (natural) vibrations when no force is applied to the system (machine), and when the pre-existing state of equilibrium is disturbed in another way. The section includes classification of mechanical vibration in machines fitted with hydraulic drive and introduction to machine as a source and receiver of mechanical vibration.

1.1.2.1 General Remarks on the Classification of Mechanical Vibrations

Operating machines and devices fitted with hydraulic systems are a source of mechanical vibrations over a wide spectrum of frequencies [16, 17]. These vibrations can generally be divided into free, forced, and self-excited categories. In this chapter, the mathematical description related to vibrations induced by a harmonic force is examined. This type of vibration will be analyzed in further chapters.

In this section, we focus on free (natural) vibrations when no force is applied to the system (machine), and when the pre-existing state of equilibrium is disturbed in another way. In working machines, such vibrations can be caused by a sudden change in the movement conditions of the machine (e.g., start-up, braking—during a sudden override of the hydraulic valves controlling the operating parameters of hydraulic systems and a sudden change in the speed of the receiver or tooling), and a sudden change in load conditions (e.g., change in the resistance of the excavated soil, the machining tool blade entering and going out of the material, and change in the direction of the load). These vibrations are of a damped nature due to resistances in the system that counteract such vibrations (damping forces). Free vibrations depend on the properties of the system and its initial state [18], i.e., on its velocity and position at the initial moment $t = 0$.

In the case of forced vibrations when the external impacts on the system (machine) are variable in time. When the external forces are periodic, the vibration forces will also be periodic. Forced vibrations can be induced kinematically, in which case we speak of kinematic forcing. In addition, vibrations can be forcefully induced, in which case we speak of force forcing. Kinematic excitations are created, for example, by transferring vibrations generated by other machines through the foundation during the movement of a mobile machine on uneven ground [19]. The causes of forced vibrations also include factors such as unbalance of rotating parts, the eccentricity of the machined surface in relation to the machined one [20] (for machine tools), and the intermittent nature of the machine operation. Some mechanical vibration impact and their applying point in machine (including hydraulic system), on the example of excavator displayed on Fig. 1.3.

Vibrations as effects of a harmonic force applied. There is an abundance of literature related to the analytical study of this type of vibration [18, 21–23]. In the case of vibration isolation, harmonic vibrations and forces are of particular importance, hence, we briefly

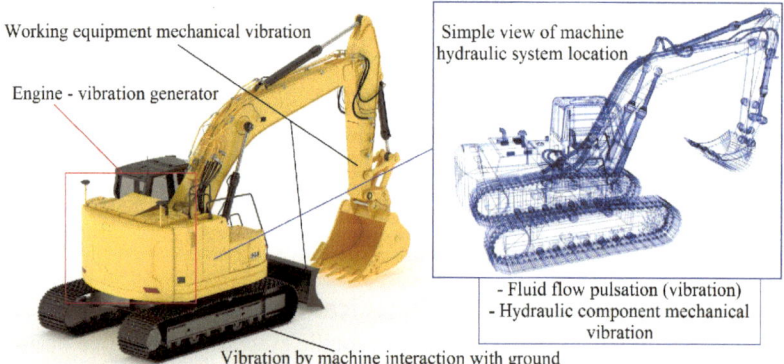

Fig. 1.3 Vibration problem in the machine—excavator vibration conditions

describe the absolute and relative vibration motion. The adoption of different reference systems results in different solutions.

The equation of motion of a system with one degree of freedom excited to move by a harmonic force is:

$$m\ddot{x}(t) + k\dot{x}(t) + cx(t) = F_0 \sin(\omega t), \tag{1.1}$$

where F_0 is the amplitude of the excitation force, and ω is the angular frequency of the excitation force, k—damping, c—stiffness, m—mass of the system, t—time.

The solution to Eq. (1.1) contains two terms [24]:

$$x(t) = e^{-\gamma \omega_0 t}(D\cos\omega_t t + E\sin\omega_t t) + \frac{F_0}{c} \frac{1}{\sqrt{\left(1 - \frac{\omega^2}{\omega_0^2}\right)^2 + \left(2\gamma \frac{\omega}{\omega_0}\right)^2}} \sin(\omega t - \Phi),$$

$$\tag{1.2}$$

where γ is the dimensionless damping coefficient, ω_t is the damped free vibration frequency, and Φ is the angle of phase delay between the excitation (force) and response (displacement), ω_0—angular frequency of undamped free vibrations, D and E—constants of integration depending on initial conditions.

The amplitude of absolute steady-state vibration x_0 is therefore described by the following formula [18, 24]:

$$x_0 = \frac{F_0}{c} \frac{1}{\sqrt{\left(1 - \frac{\omega^2}{\omega_0^2}\right)^2 + \left(2\gamma \frac{\omega}{\omega_0}\right)^2}}, \tag{1.3}$$

and the phase shift angle [18, 24]:

$$\Phi = \arctan\left(\frac{2\gamma\frac{\omega}{\omega_0}}{1 - \left(\frac{\omega}{\omega_0}\right)^2}\right). \tag{1.4}$$

Considering the minimization of the vibrations transmitted from the vibrating ground to the device or hydraulic valve, it is convenient to use kinematic forces in the analysis. If a body of mass m is subjected to harmonic kinematic forces in the form of:

$$w(t) = w_0\sin\omega t, \tag{1.5}$$

and this equation of absolute motion then takes the form of:

$$m\ddot{x}(t) + k(\dot{x}(t) - \dot{w}(t)) + c(x(t) - w(t)) = 0. \tag{1.6}$$

The solution of Eq. (1.6) for zero initial conditions takes the form of [24]:

$$\frac{x(t)}{w_0} = \frac{1}{\sqrt{\left(1 - \frac{\omega^2}{\omega_0^2}\right)^2 + \left(2\gamma\frac{\omega}{\omega_0}\right)^2}}$$

$$\left[\sin(\omega t - \Phi_1) + 2\gamma\frac{\omega}{\omega_0}\cos(\omega t - \Phi_1) - \frac{\frac{\omega}{\omega_0}}{\sqrt{1 - \gamma^2}}e^{-\gamma\omega_0 t}\sin(q_t t - \Psi)\right], \tag{1.7}$$

where the corresponding phase angles have the form of:

$$\Phi_1 = \arctan\frac{2\gamma\frac{\omega}{\omega_0}}{1 - \left(\frac{\omega}{\omega_0}\right)^2} \tag{1.8}$$

$$\Psi = \arctan\frac{2\gamma\sqrt{1 - \gamma^2}}{\frac{1}{\frac{\omega^2}{\omega_0^2}} - 1 + 2\gamma^2} \tag{1.9}$$

and q_t is a free vibration frequency damped.

After transforming Eq. (1.7), the solution can be obtained in a form convenient for the analysis of steady-state oscillations:

$$\frac{x(t)}{w_0} = \sqrt{\frac{1 + \left(2\gamma\frac{\omega}{\omega_0}\right)^2}{\left(1 - \frac{\omega^2}{\omega_0^2}\right)^2 + \left(2\gamma\frac{\omega}{\omega_0}\right)^2}} \cdot \sin(\omega t - \Phi)$$

$$- \frac{\frac{\omega}{\omega_0}}{\sqrt{1 - \gamma^2}} \cdot \frac{e^{-\gamma\omega_0 t}}{\sqrt{\left(1 - \frac{\omega^2}{\omega_0^2}\right)^2 + \left(2\gamma\frac{\omega}{\omega_0}\right)^2}} \cdot \sin(q_t t - \Psi) \tag{1.10}$$

and the phase angle is described by the following equation:

$$\Phi = \arctan \frac{2\gamma \frac{\omega}{\omega_0}}{\frac{1}{\left(\frac{\omega}{\omega_0}\right)^2} - 1 + 4\gamma^2} \tag{1.11}$$

In order to evaluate the effectiveness of absolute vibration reduction of a body of mass m, the amplification factor [24] can be introduced, which is the ratio between the amplitude of the steady-state absolute vibrations x_0 and the amplitude of the kinematic excitation w_0:

$$T_a = \frac{x_0}{w_0} = \sqrt{\frac{1 + \left(2\gamma \frac{\omega}{\omega_0}\right)^2}{\left(1 - \left(\frac{\omega}{\omega_0}\right)^2\right)^2 + \left(2\gamma \frac{\omega}{\omega_0}\right)^2}}. \tag{1.12}$$

If the frame of reference is changed, and related to the harmonically vibrating ground, then the description of the relative motion caused by the harmonic kinematic excitation takes the form of [24–26]:

$$\frac{y_0}{w_0} = \frac{\left(\frac{\omega}{\omega_0}\right)^2}{\sqrt{\left(1 - \left(\frac{\omega}{\omega_0}\right)^2\right)^2 + \left(2\gamma \frac{\omega}{\omega_0}\right)^2}} \tag{1.13}$$

The forced motion occurs with some delay relative to the force, which is expressed as [24, 26]:

$$\tan\Phi = \frac{2\gamma \frac{\omega}{\omega_0}}{1 - \left(\frac{\omega}{\omega_0}\right)^2} \tag{1.14}$$

In the case of periodic changes to the parameters of the system (e.g., its stiffness, damping, mass, and dimensions), herein, we examine parametric vibrations [27, 28]. The difference between these vibrations and, for example, self-excited vibrations, is that for self-excited vibrations, the variation of the factor that forces the vibrations is independent of the system parameters. For parametric vibrations, the values of system parameters change over time. Parametric vibrations are involved, for example, during the movement of cages in mine shafts, in which, as a result of vibrations, they change the values of elasticity coefficients [29].

Self-excited vibration (auto-oscillation) is associated with the concept of stability of motion. A stable system is considered to have no steady-state vibrations after transient excitation [30, 31]. In an unstable system, when the system is excited, vibrations usually increase until a certain cycle of vibrations with a certain waveform (shape, amplitude,

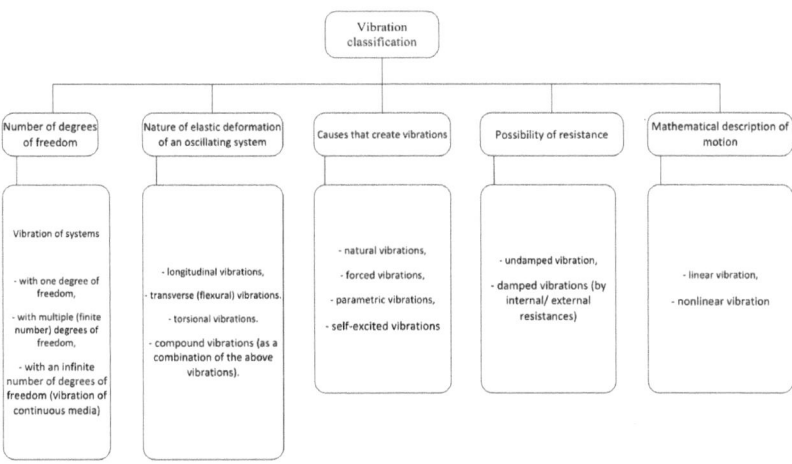

Fig. 1.4 Vibration classification

frequency) is established. This type of vibration is called self-excited vibration. Examples of self-excited vibrations include relaxation vibrations in machine tools during the movement of supports on slideways or, more generally, vibrations induced by forces caused by the motion itself. Other examples of self-excited vibrations are vibrations of icy transmission lines subjected to strong winds, vibrations of aircraft wings (flatters) [32], and vibrations of a suspension bridge (which caused the collapse of the Tacoma Strait bridge) [33, 34].

The vibrations can be divided into the following classifications (more detail displayed on Fig. 1.4):

- The number of degrees of freedom: vibrations of systems with one degree of freedom, vibrations of systems with many (finite) degrees of freedom, and vibrations of systems with an infinite number of degrees of freedom (vibrations of continuous media);
- The nature of elastic deformations of a vibrating system: longitudinal vibrations, transverse (bending) vibrations, torsional vibrations, and complex vibrations resulting from the combination of previous vibrations;
- The causes of vibrations: natural vibrations, forced vibrations, parametric vibrations, and self-excited vibrations;
- The possibility of resistance: undamped vibrations, and damped vibrations (due to the occurrence of internal resistances, e.g., intermolecular friction forces or external resistances, e.g., drag forces of the medium);
- The mathematical description of motion: linear vibrations (described by linear equations), and non-linear vibrations (described by non-linear equations).

The sources of vibrations in the environment are very diverse, and, in general, they can be divided into determined and random, or external and internal. Internal sources of vibrations are caused by the operation of machines that are set on ceilings and foundations. This type of vibration can also include vibrations caused by sanitary installation equipment, although they are usually acoustic in nature. Vibrations arising from the outside environment are transmitted through the ground, and their sources may be: Traffic (passing cars, trams), the frequency of which depends on the speed of the vehicle, and from a moving tram, the frequency of ground vibrations caused by such excitation is usually between 3 and 40 Hz [29]; Railway traffic [35]; Mmachines operating close to buildings, and factory spaces [36]; Explosions (e.g., in the excavation process of stone material in quarries), and implosions [37].

Random vibrations are caused by the forces of nature, such as winds, storms, sea waves, storms, seismic tremors, earthquakes, avalanches, etc., according to Waleed [38]. The Classification of shock sources according to [29, 39, 40] is presented in Table 1.1.

The vibrations caused by transportation are listed in Table 1.1 and are accompanied by such phenomena such as:

- The occurrence of time-varying forces at the meeting of the road surface with the vehicle wheel;
- These forces cause the road to vibrate;
- Transmission of these vibrations to the ground and their further propagation;
- Transmission of these vibrations to the vehicle structure and the neighboring objects (machines, buildings, people, machinery, and devices, in particular, those fitted with hydraulic elements and systems).

1.1.2.2 The Machine as a Source and Receiver of Mechanical Vibrations

The formation of vibrations in the vicinity of an operating machine may be stimulated by such factors as automation or mechanization, optimization of dimensions and weight, and the use of new construction materials with high wear resistance [41]. They are usually characterized by low internal damping, which reduces the ability of the machine or device to disperse energy and dampen vibrations. Also, the new requirements for machines, devices, and processes increase dynamic impacts manifested in the form of vibrations and noise (Fig. 1.5).

In many practical cases, the forces that cause vibrations act directly on the machine components or may stem from other machine or device components. In addition to the forces that vary periodically, the source of vibrations includes forces whose value is constant over time but whose direction or point of application changes (e.g., centrifugal inertial forces). In machines and devices, the source of vibrations is often kinematic excitations (e.g., when a mobile machine drives over uneven ground). Table 1.2 presents simplified diagrams of selected sources of vibrations that occur in machines and time waveforms of excitation forces.

Table 1.1 Classification of shock sources

Tremor source name	Location of tremors	Where are the sources?	How are tremors generated?	Effects of tremors: – Cause of generation, – Possibilities of prevention or protection – Possibilities of forecasting
Mining collapses	Underground, localized	Mining areas with old types of rock-burst workings, other geological areas with underground excavations	Transformation of potential energy into kinetic energy, formation of elastic waves	– Underground cavities after mining excavations, difficult or impossible forecasting
Snow and rock avalanches, volcanic activities	On the earth's surface, linearly variable	Mountain areas	Conversion and dissipation of kinetic energy into displacements, formation of elastic waves	– Forces of nature, – Avalanche barriers, – Limited forecasting
Nuclear explosions and large earthmoving explosions or missile explosions	Underground, on the surface, in the atmosphere	They do not occur without the intention of humans	Transformation of potential energy into kinetic energy, formation of elastic waves	– Dependent on humans – Prediction of effects is possible to certain limits
Blasting in quarries, mines	Localized, multi-point	Opencast mines, large construction sites	Artificial release of energy and the formation of elastic waves	– Indirectly dependent on humans – Technology of works – Limitations – Forecasting possible up to certain limits

(continued)

Table 1.1 (continued)

Impacts with the ground surface through the foundation (hammers, metallurgical pile drivers)	On the surface or shallow below	Industrial plants, steelworks	Transformation of the displacement energy into the waves	– Directly dependent on humans, – The size of the weight hitting the ground – Forecasting possible to a certain extent
Passage of rail and road vehicles: railways, trams, subways, cars, and wheeled and tracked vehicles	Surface or shallow underground linearly variable, multi-point	In the vicinity of tracks, roads, industrial plants	Conversion of displacement energy into the elastic waves	– Unevenness of the surface, movement of loads, unbalance of vehicles indirectly dependent on humans, – Possible passive vibration isolation, anti-vibration baffles – Forecasting possible with the knowledge of excitation characteristics
Vibrations of reciprocating and rotating machines and devices	On the surface through foundations,	In the vicinity of industrial plants, boiler houses, various technical devices	The kinetic energy of foundation displacement, formation of elastic waves	– Eccentricities, indirectly dependent unbalance of machines, – Passive and active vibration isolation, forecasting possible
Other sources, e.g., hurricane wind load, acoustic shock, and internal explosions	Through the foundations of the building on which the load is acting	In places exposed to these influences	Dynamic loads transferred to the foundations cause the formation of elastic waves	– Forces of nature, independent or indirectly dependent, – The possibility of isolation only in exceptional cases – Forecasting impossible

The coincidence of mechanical vibrations and pressure pulsations in hydraulic systems can be considered in two ways. Firstly, the hydraulic system components are induced to vibrate by the pulsating flow (e.g., vibrations of pipes, hoses, or valves), and secondly, pressure pulsation is induced by vibrations of hydraulic system elements, and in particular, valve control elements are excited (e.g., spools of directional valves or poppets of maximum valves).

In a hydraulic system, variation of fluid pressure can be due to pulsation of the working liquid capacity, which stems from the kinematics of the pump's displacement elements. Pulsation of the pump efficiency leads to periodic changes in pressures owing to the

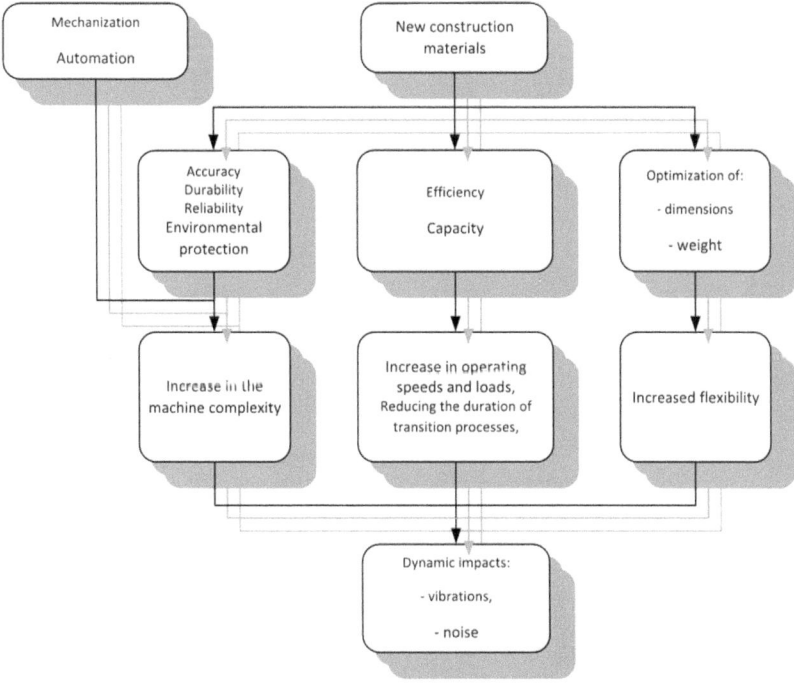

Fig. 1.5 Classification of reasons for the increase of dynamic interactions in machines

harmonic function in the hydraulic system. The frequency of this pressure pulsation corresponds to the frequency of the capacity pulsation [42]. In addition, the pulsation excites mechanical vibrations, e.g., hydraulic lines. An example of the excitation of such vibrations is shown in Fig. 1.6 and is based on the authors' measurements.

As shown in the spectrum (Fig. 1.6), the harmonic components of the mechanical vibrations corresponding to the pressure pulsation can be observed by the pressure pulsation of the flowing liquid caused by the mechanical vibrations of the hydraulic line.

Identification of the impact of pressure pulsations on the elements of the hydraulic system is achieved by a specially designed test stand, an important element of which is the latest generation single-stage proportional valve with the DFPlus symbol. The stand enables the generation of pressure pulsations in the frequency range of up to 350 Hz (the cut-off frequency of the DFPlus valve). Figure 1.7 shows the diagram of the stand's hydraulic system.

The following parameters can be measured and recorded: the waveform of the control signal of the DFPlus directional valve, displacement of the DFPlus valve spool, waveforms of pressures upstream and downstream of the tested 4WRE valve, vibration acceleration of the body of the tested 4WRE valve, and displacement of the spool of the tested 4WRE

Table 1.2 Selected diagrams of vibration sources in machines and time waveforms of excitation forces

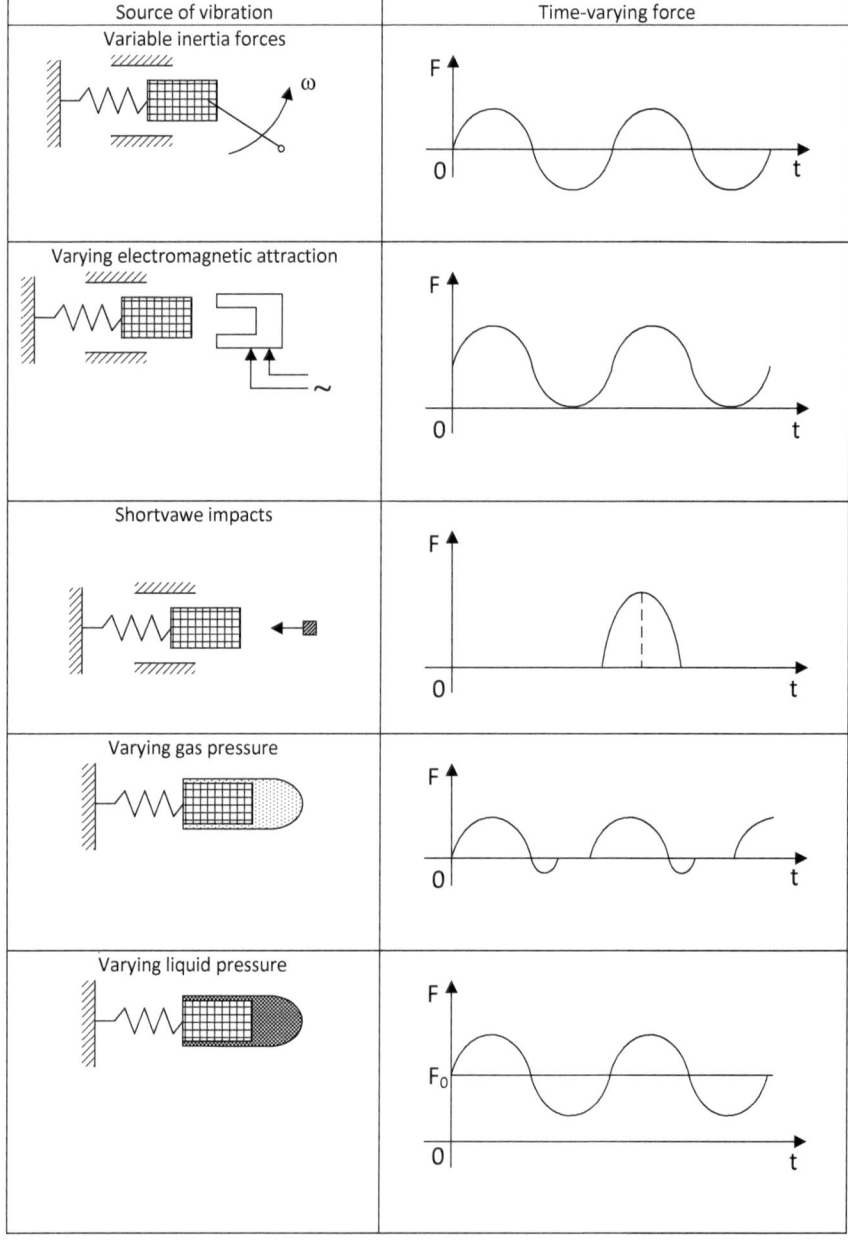

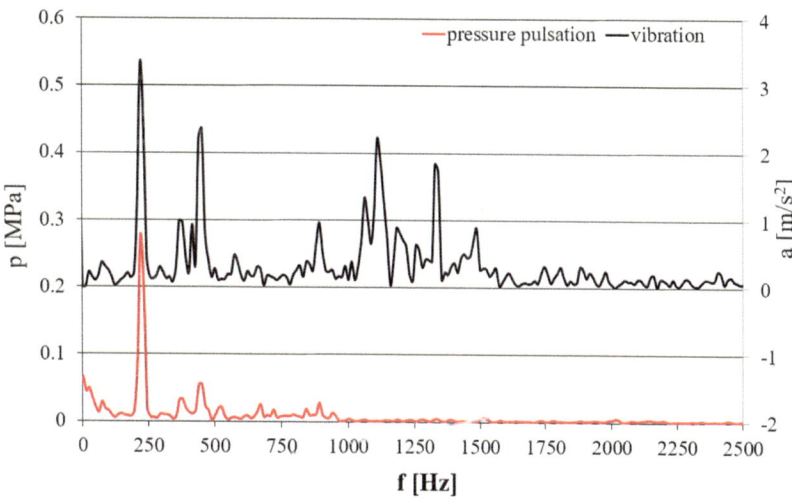

Fig. 1.6 The amplitude-and-frequency spectrum of pressure pulsations and mechanical vibrations of a hydraulic line $p_{avrage} = 5$ MPa; $Q = 1.83 \cdot 10^{-5} m^3/s$ (1.1 dm³/min)

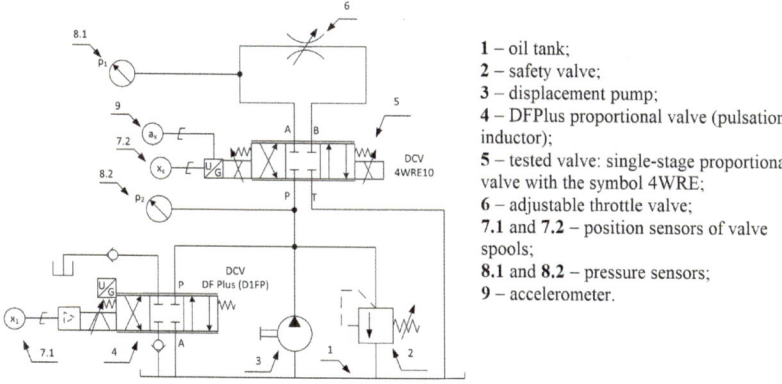

1 – oil tank;
2 – safety valve;
3 – displacement pump;
4 – DFPlus proportional valve (pulsation inductor);
5 – tested valve: single-stage proportional valve with the symbol 4WRE;
6 – adjustable throttle valve;
7.1 and 7.2 – position sensors of valve spools;
8.1 and 8.2 – pressure sensors;
9 – accelerometer.

Fig. 1.7 Diagram of the hydraulic system of the test stand to investigate the effects of pressure pulsations on the hydraulic valve

directional valve. The diagram of the control, measurement, and data acquisition system is shown in Fig. 1.8.

The software, in cooperation with the stand, enabled the generation of the waveform of a signal controlling the DFPlus valve independent of time. The valve control procedure using the above-mentioned software is described detail in Kudźma [43]. A harmonic signal with a given frequency f and amplitude s_0, and a constant shift relative to zero s_{01} of the following form is used $s = s_{01} + s_0 \cdot \sin(2\pi f t)$. The test results that display

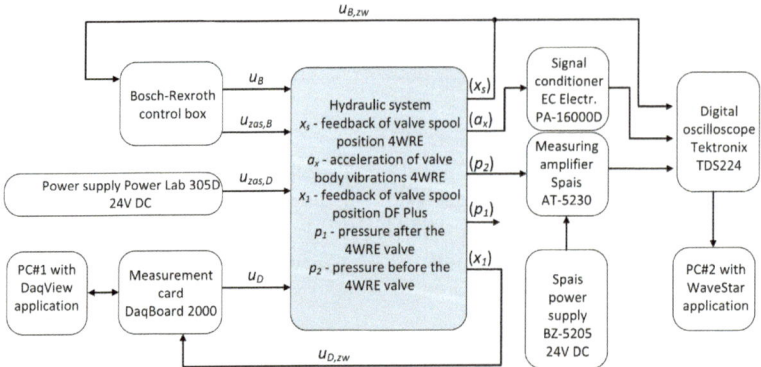

Fig. 1.8 The control, measurement, and data acquisition system of the test stand for the generation of pressure pulsation

the excitation of the mechanical vibrations of the hydraulic directional valve body as a result of pressure pulsations are shown in Fig. 1.9, with the following control signal parameter values: $s_{01} = 5$ V, $s_0 = 4$ V. A modular design of the testing stand allows for the replacement of the 4WRE valve (position 5 on the Fig. 1.7) with another hydraulic valve.

In the showed spectrum of pressure pulsations (Fig. 1.9) upstream of the tested 4WRE proportional valve at point 8.2 (Fig. 1.7) and the acceleration of vibrations of the directional valve body at point 9 in the axis of the spool. The results confirm the correlation between the pressure pulsation in the hydraulic system and the vibrations of the valves to the fitted system.

The components of the hydraulic systems in which machines and devices are fitted are subjected to vibrations. The effects of these vibrations largely depend on their direction, frequency, amplitude, and duration [44]. Therefore, in order to identify the excitations that cause the mechanical vibrations, reported analysis can be employed of the acceleration of vibrations of the hydraulic power-pack tank (Fig. 1.10a, b) that involve the direction of movement of the hydraulic valve control element mounted on the tank.

The reported analysis of the vibration acceleration of the forklift frame shown in Fig. 1.11a, b indicate that the excitation spectrum also includes lower frequency components.

The reconnaissance tests have been conducted on the Ł220 loader, whose frame (in the vicinity of the rear wheel arch) vibration acceleration was measured with a triaxial accelerometer for driving forward without load. The results are presented in the form of an average 1/3 octave spectrum in Fig. 1.12a. Figure 1.12b spectrum is limited to 80 Hz, and the measuring point is located on the frame of the machine near the drive motor. In figures present a wide spectrum of forces applied to the loader frame, which include components with frequencies below 100 Hz.

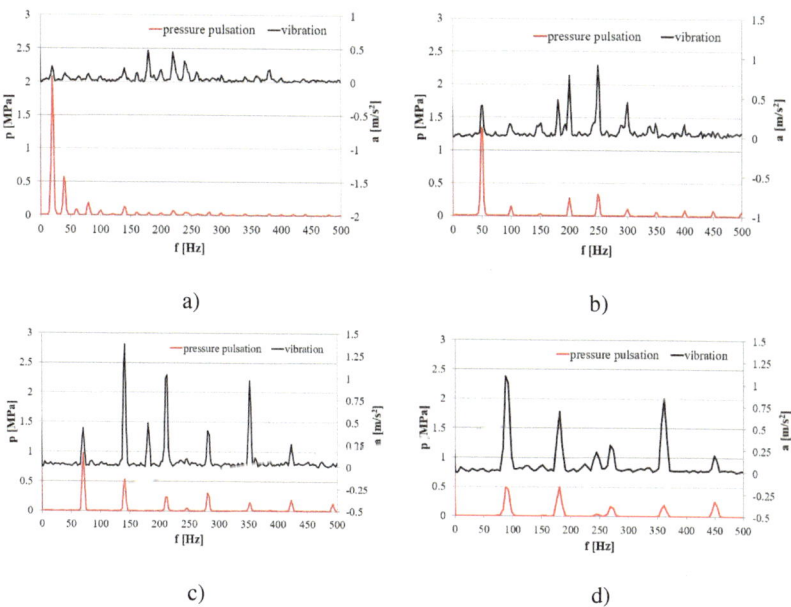

Fig. 1.9 The amplitude-and-frequency spectra of pressure pulsations and excited mechanical vibrations of the body of the 4WRE proportional valve ($p_{average} = 2.5$ MPa; $Q = 1.083 \cdot 10^{-4}$ m^3/s (6.5 dm^3/min))—frequency of the control signal of the DFPlus valve: **a** $f = 20$ Hz; **b** $f = 50$ Hz; **c** $f = 70$ Hz; **d** $f = 90$ Hz

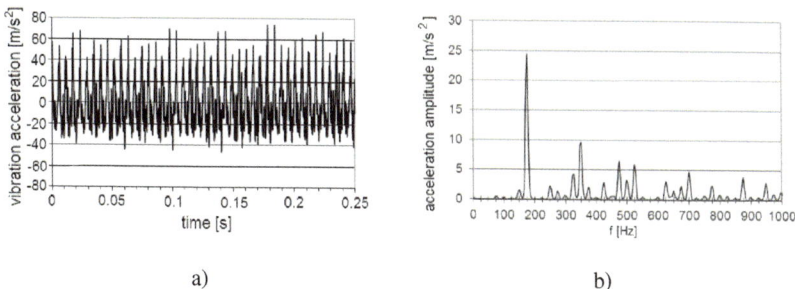

Fig. 1.10 Mechanical vibration measurements—the rotational speed of the pump shaft $n = 1450$ rev/min. (Authors' measurements): **a** Recorded acceleration of the plate's vibration of the hydraulic power-pack tank; **b** The amplitude-and-frequency spectrum of the plate's vibration acceleration of the hydraulic power-pack tank

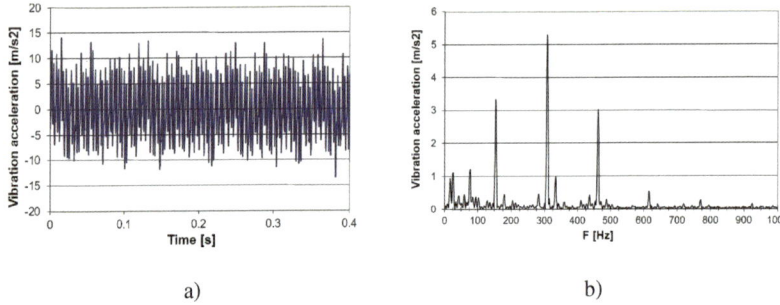

a)   b)

Fig. 1.11 Mechanical vibration measurements on forklift—drive motor rotational speed 800 rev./min. (Authors' measurements): **a** Recorded vibration acceleration of the plate on which the hydraulic control elements of the forklift are mounted; **b** The amplitude-and-frequency spectrum of vibration acceleration of the forklift mounting plate

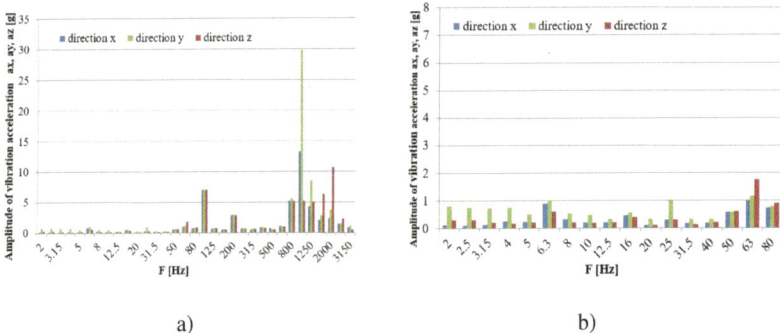

a)   b)

Fig. 1.12 The average 1/3 octave spectrum of the vibration acceleration (Authors' measurements): **a** The rear frame of the Ł220 loader; **b** The frame of the construction loader "Ł-220" from Bumar-Fadroma (Spectrum limited to 80 Hz)

Additionally, other vehicles are a source of mechanical vibrations in which the spectrum includes low frequencies—Tables 1.3 and 1.4, based on the [45, 46].

The vibration test results of selected vehicles and devices (Figs. 1.10b, 1.11b and 1.12) clearly indicate the presence of vibrations in these vehicles with a frequency of several to several hundred Hz and an amplitude close to 2000 m/s^2.

The literature indicates that machines fitted with hydraulic systems are a source of mechanical vibrations with a wide spectrum of frequencies and significant amplitudes. They affect both humans and the components of hydraulic systems. In addition, the elements of the hydraulic system in which the machine is fitted with a receiver of mechanical vibrations can disturb their proper operation.

Table 1.3 Results of vibration tests of selected vehicles (1/2)

Type of vibrations and vibrating system	Frequency [Hz]	Maximum acceleration [g]
Car body on springs	2–5	8
Axles with wheels on tires	12–17	2.5
Car body as a deformable system	60–120	1
Impacts of wagons when being connected (horizontal vibrations)	6–9	14.6
Vertical vibrations of a loaded wagon at a speed of 50 km/h	3–8	3
Vibrations from impacts against the rails where they are joined (vertical vibrations measured in the wagon)	8–30	0.1–2.2
Vibrations from impacts affecting the equipment and apparatus on board the ship	0–12	30–200

Table 1.4 Results of vibration tests of selected vehicles (2/2)

Type of vehicle and place of measurement	Frequency [Hz]	Acceleration [g]	
		Medium	Maximum
Tractor			
Front axle (vertical vibration)	9–10	2.5	5
Chassis frame excited by engine vibration	60–100	0.07	0.25
Truck			
Unladen on a concrete-surfaced road at 60 km/h	5–15	0.29	1.5
Tank			
Vertical vibrations in the plane of the driver's seat and during off-road movement	4–8	0.7	12.5

1.1.3   Fluid Pulsation (Vibrations) in Machines Hydraulic Systems

Improvement the stability of hydraulic machines, as well as increase of their efficiency is impossible without analysising hydrodynamic processes, where an important
role is played by transient phenomena associated with the development of pulsation
and vortex structures of fluid flow. Causes of flow pulsation inside the hydraulic drive
and its influence on main connection components—pipelines, are presented in current
sub-chapter.

1.1.3.1   Flow Pulsation and Damping in Machines Hydraulic Systems

Flow in hydraulic drives is associated with a variety of transient phenomena, according
to Minakov et al. [47] and Stosiak et al. [48], which can be divided into three groups by
pulsation:

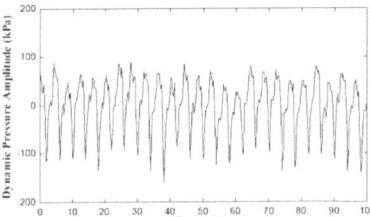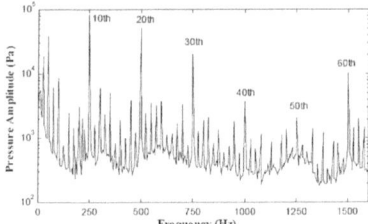

Fig. 1.13 An investigation into the pulsation of a hydraulic pump [50]: **a** pressure waveform at the pump outlet (1500 rpm); **b** frequency of the pressure waveform (main 25 Hz)

- High-frequency range—hydraulic pumps;
- Middle-frequency range—hydraulic components (including valve etc.);
- Low-frequency range—excited in the pipelines.

Pulsations arise as a consequence of the emergence of distinct vortices when alterations are made to the cross-sectional area within the hydraulic drive lines. In the realm of mobile hydraulic drives, variable displacement pumps are commonly employed to cater to specific volume flow requirements, as noted by Van Basshuysen et al. [49]. However, due to the inherent nature of positive displacement pumps, they generate a pulsatile flow, causing fluctuations in pressure amplitudes, as depicted in Fig. 1.13 [50].

As highlighted by Baum and Erzberger [51], hydraulic pumps often serve as the primary source of noise and vibrations within a hydraulic system. Additionally, Hoffmann [52] and Beynart [53] have emphasized that these pulsations can lead to increased dynamic loads, wear, and higher energy consumption throughout the hydraulic system. To mitigate the adverse effects of pulsations, the integration of pulsation dampers downstream of the pump is widely employed in hydraulic drives. These dampers can be incorporated either as a side branch in the pipeline system, as mentioned by Koegler et al. [54], or integrated directly into the pump outlet. However, designing pulsation dampers poses significant challenges, especially when the pump is driven by a variable-speed combustion engine. In such cases, the dampers must operate reliably over a broad range of frequencies and working pressures, as emphasized by Chai et al. [55]. This requirement makes the task of ensuring effective pulsation control even more demanding.

The result of any damping is usually expressed in decibels (dB). Typical damping factors for hydraulic dampers range from 10 to 40 dB or even higher, as reported by Bach et al. [56]. These damping factors represent the level of attenuation the damper provides to the pulsations at a specific frequency, helping to reduce noise and vibrations in the hydraulic system, can be calculated using the following equation:

$$\text{Damp} = 20 \cdot \log\left(\frac{p_0}{p_1}\right)(dB), \tag{1.15}$$

where p_0 and p_1—amplitude of the pressure pulsations of the baseline measurement without a damper and with a damper respectively.

The first detail research by Esser [57] classifies pulsation dampers into two categories: interference and absorption type dampers. The absorption principle involves dissipating a portion of the pulse energy into heat through fluid friction or internal material friction, as explained by Hoffmann [52]. On the other hand, the interference or reflection principle relies on destructive interference, achieved by superposing waves with a $180°$ phase difference induced by changes in pipeline diameter, as described by Mikota [58].

Pulsation dampers, much like hydraulic accumulators, are often quite large, making integration into certain mobile machine's hydraulic systems challenging. Moreover, hydraulic accumulators may not function effectively across the required frequency range, or they may impose significant pressure issues on the system, as demonstrated by Kitajima et al. [59]. To address these challenges, some hydraulic systems incorporate pumps with integrated pulsation dampers suitable for mobile applications. However, integrating dampers into pumps can be problematic due to the space requirements.

Nevertheless, achieving accurate dynamic analysis is often difficult because hydraulic driving processes are strongly nonlinear due to the compressibility of oil and the nonlinear characteristics of both the pump and the valve with hydraulic pipeline combination.

1.1.3.2 Influence of Fluid Pulsation Inside Hydraulic Systems on the Pipelines

In a hydraulic drive system, the pipeline comprises several different layers, including rubber and metal braid, as depicted in Fig. 1.14. The construction of the pipeline involves combining these layers to provide strength, flexibility, and durability. The innermost layer is typically made of rubber, which serves as the primary fluid-carrying component. Rubber is chosen for its resistance to hydraulic fluids and its ability to maintain a leak-free seal. Surrounding the rubber layer, there is a metal braid. This braided reinforcement adds structural integrity to the pipeline and helps to prevent the rubber from expanding under high pressure. The metal braid enhances the pipeline's resistance to external forces, such as mechanical impacts or vibrations. The combination of rubber and metal braid provides a balanced pipeline that can handle the high pressures and dynamic conditions often present in hydraulic systems. The rubber ensures fluid compatibility and sealing properties, while metal braid reinforces the pipeline to maintain its shape and mechanical strength. This layered design is essential for the efficient and reliable functioning of hydraulic drives.

In the review conducted by Paidoussis and Li [60], the influence of fluid pulsation inside pipelines was extensively explored through an examination of more than a hundred articles covering various aspects of the issue. Wang and Ni [61] have observed that fluid flow models within pipelines have been formulated in stable or unstable forms. Paidoussis [62] have highlighted that one of the primary causes of fluid oscillations in pipelines is the continuously varying flow velocity over time. These periodic pulsations in the flow

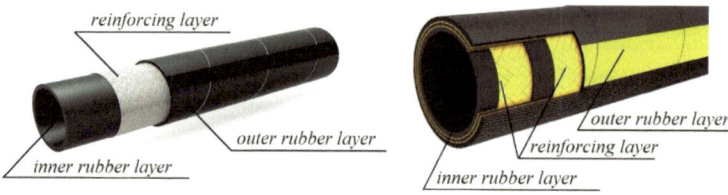

Fig. 1.14 Metal-braided hydraulic pipelines construction

are linked to the operating characteristics of the feed pump or the established flow rate control. The flow pulsations, resulting from pump operation, manifest as a polyharmonic signal with one dominant component, leading to considerable pressure and power losses in the hydraulic system, as well as continuous vibrations in the pipelines, according to the studies by Fornarelli et al. [63]. These pulsations can have adverse effects on the overall performance and reliability of the hydraulic system.

In the Fig. 1.15, presented by Wang [64], illustrates the displacement velocity of a pipeline structure from one of its layers, as well as its corresponding frequency response. The graph provides insights into the dynamic behavior of the pipeline at different frequencies, showcasing how the structure responds to the applied forces or vibrations.

The displacement velocity indicates the speed at which the pipeline layer moves in response to external stimuli, such as fluid pulsations or mechanical disturbances. The frequency response curve demonstrates how the displacement velocity varies across a range of frequencies, revealing the pipeline's natural frequencies and resonant behavior. This information is crucial for understanding the pipeline's dynamic characteristics, allowing engineers and researchers to design and optimize the pipeline's construction to mitigate potential issues related to vibrations, fatigue, and other mechanical concerns. Chen [65]

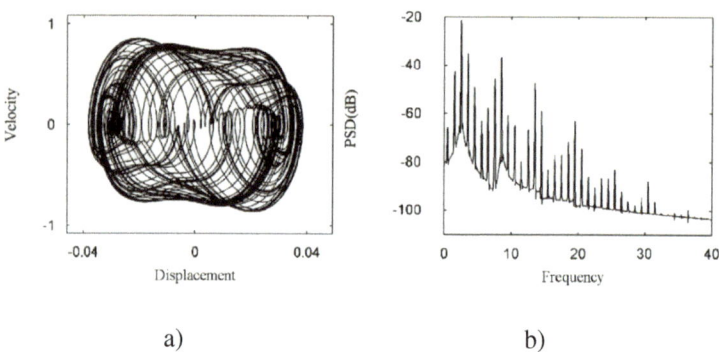

a)   b)

Fig. 1.15 One-layer pipeline deformation [64]: **a** velocity displacement of pipeline structure; **b** frequency response of velocity

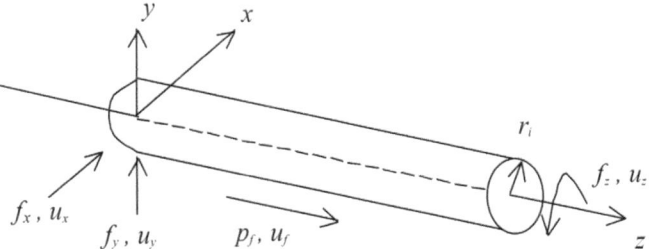

Fig. 1.16 Hydraulic line simulation [65]: u—translational displacement of the pipeline; f_x, f_y, f_z— forces acting on the tubing; p_f—fluid pressure; u_f—fluid displacement

introduced a mathematical model to investigate the influence of fluid flow on pipeline deformation. The model focuses on the deformation of a single-layer pipeline caused by the movement of fluid inside it. To facilitate the analysis, Fig. 1.16 illustrates the coordinate system and state variables within a straight tube segment.

The coordinate system likely defines the spatial reference frame used to describe the deformation and fluid flow behavior. It could involve position coordinates and directions within the tube segment. State variables refer to the parameters that describe the system's current state at a specific time. In this context, they may represent variables such as fluid pressure, flow velocity, and the deformation or displacement of the pipeline at different points along the tube segment. By utilizing this mathematical model and the specified coordinate system and state variables, researchers can gain insights into the fluid-induced deformation of the pipeline and analyze how various fluid flow conditions impact its structural integrity and performance. This analysis is crucial for understanding the mechanical behavior of pipelines under fluid flow and optimizing their design and operation to ensure reliable and safe performance.

The pressure variation along the length of a pipeline is determined by a specific mathematical function Chen [65]:

$$\frac{\partial P}{\partial z} = \frac{p_f s^2}{\Omega(s)} U_f - \rho_f \left[\frac{2 + \Omega(s)}{\Omega(s)} \right] U_z; \tag{1.16}$$

$$\Omega(s) = \left[\frac{2 J_1 \left(j r_i \sqrt{s/v} \right)}{J_0 r_i \sqrt{s/v} J_i \left(J_1 r_i \sqrt{s/v} \right)} - 1, \right] \tag{1.17}$$

where P—fluid pressure; ρ_f—fluid density; J_0 and J_1—zero- and first-order Bessel functions of the first kind; U_f—fluid displacement; U_z—displacement of pipeline inner surface; v—derivation of axial velocity; s—denotes the Laplace transformation; r_i—radius of pipeline; z—coordinate.

Equations were derived by researcher Chen [65] after applying Newton's second law to both the fluid flow and the tubing wall:

$$F_t + \frac{\partial F_z}{\partial z} = \rho A s^2 U_{z;} \tag{1.18}$$

$$-A_f \frac{\partial P}{\partial z} - F_t = \rho_f A_f s^2 U_f. \tag{1.19}$$

where F_t—friction force per unit length acting on the inner pipeline wall; F_z—force per unit length acting along z coordinate; A—cross-section area of the pipeline; ρ—density of pipeline material; A_f—cross-section area of fluid.

For a pipeline, a wall can be represented by radial and tangential stresses as:

$$\sigma_r + \sigma_\theta = \frac{2\rho_p r_i^2}{r_0^2 - r_i^2,} \tag{1.20}$$

where ρ_p—density of pipeline material.

In recent research in this field, various papers have investigated nonlinear models of flexible pipelines with fixed system parameter values. For instance, Gorman et al. [66] and Luczko and Czerwinski [67] have examined such models. Additionally, some studie, like Dai et al. [68], have utilized bifurcation diagrams to illustrate the impact of a single parameter on oscillation characteristics. The primary focus of this research area lies in understanding how materials deform within pipelines under fluid pulsation and its influence on the damping of fluid pulsation. However, conducting an accurate dynamic analysis presents challenges due to several factors. The deformation force in pipelines is complex due to the nonlinear relationships between material properties, stress, stress ratio, temperature, and the irregular fluid pulsation as identified by Lin [69]. The friction force between different layers of hydraulic pipeline is also inevitable and highly nonlinear, especially during high-velocity and pressure operations.

A significant complexity arises from the dynamic coupling between the mechanical and hydraulic systems, where motion and force mutually transfer between fluid flow and pipeline material. This intricate interaction adds to the difficulty of achieving an accurate analysis in this research domain. To investigate how uses of different pipelines influence for reducing vibration in the hydraulic system by fluid flow pulsation can be used frequency response method 1973 [70]. An excellent analysis of stress–strain and frequency response in reinforced pipelines done by Van der Horn and Kuipers [71] and an American standard for testing pipelines for automotive hydraulic brakes ANSI/ASTM D5571-76 but it includes only experimental and natural (modal) frequency static tests. The different moving modal characteristics of the hydraulic pipelien at different frequency according to experimental works of Czerwinski and Luczko [72] is show in Fig. 1.17.

The central issue lies in the prevalent usage of linear models by researchers. While linear models with distributed parameters are adept at calculating the natural frequency characteristics of hydraulic pipelines without considering fluid influence, they fall short in capturing the complete frequency characteristics of fluid-mechanical power systems. To achieve a more accurate representation, the adoption of non-linear models for component

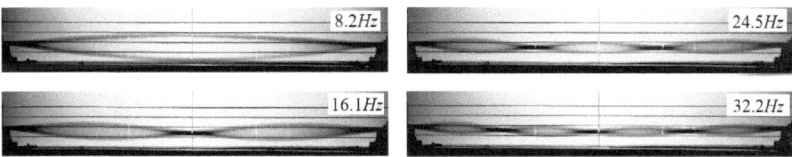

Fig. 1.17 Hydraulic pipelien vibrations modes [72]

characteristics is essential. Incorporating non-linear models for components necessitates the calculation of frequency characteristics through transient responses, as elucidated by Xu et al. [73]. In the realm of computing dynamic transient responses for fluid power systems, it is prudent to opt for simplified non-linear models with lumped parameters for hydraulic transmission lines. This choice not only conserves computational resources but also facilitates the investigation of the intricate interplay between fluid pulsation within high-pressure hoses and the frequency variations in hydraulic components. A model featuring concentrated parameters offers a valuable avenue for exploring the relationship between fluid pulsation within pipelines and the frequency fluctuations in hydraulic components. This model holds promise for enhancing the modeling and simulation of energy-efficient liquid power systems.

1.1.4 Objectives and Main Content of the Book

The subject of this book is to examine the influence of mechanical vibration on the changes in the pressure pulsation spectrum of hydraulic systems. We show that machines and equipment equipped with hydraulic systems are a source of vibration with a wide frequency spectrum. Additionally, hydraulic valves are also exposed to vibration. Vibrations of the substrate on which the hydraulic valve is installed force the control element of the hydraulic valve to vibrate. The control element's vibration produced in this way causes changes in the pressure pulsation spectrum of the hydraulic system. A friction model modified using mixed friction theory can be used for the oscillating motion of the hydraulic directional control spool. Passive vibration isolation methods are proposed to reduce valve vibration. Numerical methods are employed to analyze the effect of changes in the pressure pulsation spectrum on the hydraulic efficiency of the lines. Examples are provided for the implementation of numerical methods in the calculation of hydraulic components and systems.

References

1. Commission, E. (2011). *Transport 2050.*
2. Helbig, A. (2002). Injection moulding machine with electric-hydrostatic drives. In *3rd International Fluid Power Conference (3. IFK)* (pp. 67–82).
3. Karpenko, M., Bogdevičius, M. (2017). Review of energy-saving technologies in modern hydraulic drives. *Science—Future of Lithuania—Mokslas—Lietuvos Ateitis, 9*(5), 553–558 (VGTU Press). ISSN 2029-2341. eISSN 2029-2252. https://doi.org/10.3846/mla.2017.1074
4. Borghia, M.; Zardina, B.; Pintorea, F.; Belluzzia, F. 2014. Energy savings in the hy-draulic circuit of agricultural tractors. In *68th Conference of the Italian Thermal Machines Engineering Association, Energy Procedia* (Vol. 45, pp. 352–361). https://doi.org/10.1016/j.egypro.2014.01.038
5. Adachi, J., Siebrits, E., Pierce, A., Desroches, J. (2007). Computer simulation of hy-draulic fractures. *International Journal of Rock Mechanics and Mining Science, 44*, 739–757. https://doi.org/10.1016/j.ijrmms.2006.11.006
6. Wang, X., Huang, Z., & He, X. (2010). Diagnosis of vibration in the hydraulic system for rolling mill. *Southern Metals, 3*, 8–10.
7. Stosiak, M. (2015). Ways of reducing the impact of mechanical vibrations on hydrau-lic valves. *Archives of Civil and Mechanical Engineering, 15*(2), 392–400. https://doi.org/10.1016/j.acme.2014.06.003
8. Ortwig, H. (2005). Experimental and analytical vibration analysis in fluid power sys-tems. *International Journal of Solids and Structures, 42*, 5821–5830. https://doi.org/10.1016/j.ijsolstr.2005.03.028
9. Han, Q., & Wang, H. (2010). Prevention and improvement measures for the vibration and noise of the hydraulic system of a numerical control machine. *Mechanical Science and Technology for Aerospace Engineering, 29*, 1110–1121.
10. Amirante, R., Distaso, E., & Tamburrano, P. (2014). Experimental and numerical analysis of cavitation in hydraulic proportional directional valves. *Energy Conversion and Management, 87*, 208–219. https://doi.org/10.1016/j.enconman.2014.07.031
11. Makaryants, G., Andrey, B., Prokofiev, V., Evgeny, V., Shakhmatov, A. (2015). Vi-broacoustics analysis of punching machine hydraulic piping. In *Dynamics and Vi-broacoustics of Machines (DVM2014), Procedia Engineering* (Vol. 106, pp. 17–26). https://doi.org/10.1016/j.proeng.2015.06.004
12. Yakushev, A. (2005). Investigation of energy-saving systems. *Construction and Road Machines, 12*, 35–38.
13. Nesmiyanov, I., & Khavronin, V. (2007). Elastic drive hydraulic pump as a way to re-duce energy consumption of hydraulic machines. *Tractors and Agricultural Machinery, 6*, 45–56.
14. Shen, W., Jiang, J., Su, X., & Karimi, H. (2015). Control strategy analysis of the hydrau-lic hybrid excavator. *Journal of the Franklin Institute, 352*(2), 541–561. https://doi.org/10.1016/j.jfranklin.2014.04.007
15. Chenxiao, N., Xushe, Z. (2012). Study on vibration and noise for the hydraulic system of hydraulic hoist. In *Proceedings of 2012 International Conference on Mechanical Engineering and Material Science (MEMS)* (pp. 126–128). https://doi.org/10.2991/mems.2012.95
16. Miao, H., Wang, C., Li, C., et al. (2023). Nonlinear dynamic modeling and vibration analysis of whole machine tool. *International Journal of Mechanical Sciences, 245*, 108122. ISSN 0020-7403. https://doi.org/10.1016/j.ijmecsci.2023.108122

17. Khajehdezfuly, A., Shiraz, A., Sadeghi, J. (2023). Assessment of vibrations caused by simultaneous passage of road and railway vehicles. *Applied Acoustics, 211*, 109510. ISSN 0003-682X. https://doi.org/10.1016/j.apacoust.2023.109510

18. Harris, C., Piersol, A. (2009). *Harris' shock and vibration handbook* (6th ed.). McGraw Hill. ISBN:10-0071508198.

19. Ren, Y., Qu, S., Yang, J., et al. (2023). An efficient three-dimensional dynamic stiffness-based model for predicting subway train-induced building vibrations. *Journal of Building Engineering, 76*, 107239. ISSN 2352-7102. https://doi.org/10.1016/j.jobe.2023.107239

20. Joshi, P. (2007). *Machine tools handbook* (1st ed.). McGraw Hill. ISBN:10-0071494359.

21. Yang, K., Hu, Y., Wu, H., et al. (2023). Harmonic vibration suppression of maglev rotor system under variable rotational speed without speed measurement. *Mechatronics, 91*, 102956. ISSN 0957-4158. https://doi.org/10.1016/j.mechatronics.2023.102956

22. Inman, D.J. (2013). *Engineering vibrations*. Pearson.

23. Benaroya, H., Nagurka, M., Han, S. (2022). *Mechanical vibration*. Rutgers University Press. 1978831064.

24. Wen, B., Huang, X., Li, Y., & Zhang, Y. (2022). *Vibration utilization engineering* (p. 351). Springer.

25. Munchhof, M. (2014). *Identification of dynamic systems*. Springer. 9783642422676.

26. Wereley, N. (2021). *Introduction to vibration in engineering*. Cognella Academic Publishing. 1793563314.

27. Xu, C., Wang, Z., Zhang, H., et al. (2022). Investigation on mode-coupling parametric vibrations and instability of spillway radial gates under hydrodynamic excitation. *Applied Mathematical Modelling, 106*, 715–741. ISSN 0307-904X, https://doi.org/10.1016/j.apm.2022.02.013

28. Lu, Q., Sun, Z., Zhang, W. (2020). Nonlinear parametric vibration with different orders of small parameters for stayed cables. *Engineering Structures, 224*, 111198. ISSN 0141-0296. https://doi.org/10.1016/j.engstruct.2020.111198

29. Schmitz, T., Smith, K. (2021). *Mechanical vibrations. Modeling and measurement*. Springer. https://doi.org/10.1007/978-3-030-52344-2

30. Zhu, H., Zhang, L., Chen, Q., et al. (2023). Theoretical and experimental study on the self-excited vibration of a flexible rotor system with floating spline. *Chinese Journal of Aeronautics*. ISSN 1000-9361. https://doi.org/10.1016/j.cja.2023.03.030

31. Shaomin, L., Haichun, P., Chunjian, L., et al. (2022). Nonlinear characteristic and chip breaking mechanism for an axial low-frequency self-excited vibration drilling robot. *International Journal of Mechanical Sciences, 230*, 107561. ISSN 0020-7403. https://doi.org/10.1016/j.ijmecsci.2022.107561

32. Prakash, S., Kumar, R., Raja, S., et al. (2016). Active vibration control of a full scale aircraft wing using a reconfigurable controller. *Journal of Sound and Vibration, 361*, 32–49. ISSN 0022-460X. https://doi.org/10.1016/j.jsv.2015.09.010

33. Wu, W., Ju, J., Zhang, J., et al. (2023). Vibration phase difference analysis of long-span suspension bridge during flutter. *Engineering Structures, 276*, 115351. ISSN 0141-0296. https://doi.org/10.1016/j.engstruct.2022.115351

34. Liu, Z., Chen, L., Sun, L., et al. (2023a). Multimode damping optimization of a long-span suspension bridge with damped outriggers for suppressing vortex-induced vibrations. *Engineering Structures, 286*, 115959. ISSN 0141-0296. https://doi.org/10.1016/j.engstruct.2023.115959

35. Liu, W., Liang, R., Zhang, H., et al. (2023b). Deep learning based identification and uncertainty analysis of metro train induced ground-borne vibration. *Mechanical Systems and Signal Processing, 189*, 110062. ISSN 0888-3270. https://doi.org/10.1016/j.ymssp.2022.110062

36. Wang, S., Zhu, S. 2022. Vibration impact of rock excavation on nearby sensitive buildings: An assessment framework. *Soil Dynamics and Earthquake Engineering, 163*, 107508. ISSN 0267-7261. https://doi.org/10.1016/j.soildyn.2022.107508

37. Cheng, R.; Chen, W.; Hao, H.; Li, J. 2023. Numerical prediction of ground vibrations induced by LPG boiling liquid expansion vapour explosion (BLEVE) inside a road tunnel. *Underground Space, 12*, 44–64. ISSN 2467–9674. https://doi.org/10.1016/j.undsp.2023.02.007

38. Waleed, F. (2011). *Random vibration of non-conformal contact systems*. Lambert Academic Publishing.

39. Griffin, M. (1990). *Handbook of human vibration*. Academic Press. https://doi.org/10.1016/C2009-0-02730-5

40. Li, A. (2020). *Vibration control for building structures. Theory and applications*. Springer. https://doi.org/10.1007/978-3-030-40790-2

41. Novillo, E. (2021). *Vibration control engineering*. Taylor & Francis Ltd., 1032006994.

42. Xu, W., Wang, Z., Zhou, Z., et al. (2023). An advanced pressure pulsation model for external gear pump. *Mechanical Systems and Signal Processing, 187*, 109943. ISSN 0888-3270. https://doi.org/10.1016/j.ymssp.2022.109943

43. Kudźma, Z. (2012). Tłumienie pulsacji ciśnienia i hałasu w układach hydraulicznych w stanach przejściowych i ustalonych. Oficyna Wydawnicza PWr. (In Polish).

44. Yang, M., Gui, L., Hu, Y., et al. (2018). Dynamic analysis and vibration testing of CFRP driveline system used in heavy-duty machine tool. *Results in Physics, 8*, 1110–1118. ISSN 2211-3797. https://doi.org/10.1016/j.rinp.2018.01.067

45. Rogers, N. (2022). Transport noise. In N. Rogers (Ed.). The *Maltings West Street Bourne Lincs PE10 9PH* (pp. 0308–437). Warners Group Publications.

46. Fan, Y., Teo, H. P., & Wan, W. X. (2021). Public transport, noise complaints, and housing: Evidence from sentiment analysis in Singapore. *Journal of Regional Science, 61*, 570–596. https://doi.org/10.1111/jors.12524

47. Minakov, A., Platonov, D., Dekterev, A., Sentyabov, A., & Zakharov, A. (2015). The analysis of unsteady flow structure and low frequency pressure pulsations in the high-head Francis turbines. *International Journal of Heat and Fluid Flow, 53*, 183–194. https://doi.org/10.1016/j.ijheatfluidflow.2015.04.001

48. Stosiak, M., Karpenko, M., Prentkovskis, O., Deptuła, A., Skačkauskas, P. (2023). Research of vibrations effect on hydraulic valves in military vehicles. In *Defence technology* (pp. 1–15). KeAi Publishing LTD. ISSN 2096-3459. In Press.

49. Van Basshuysen, R., Schäfer, F. (2017). *Handbuch verbrennungsmotor: Grundlagen, komponenten, systeme, perspektiven* (p. 1385). Springer. (in German). https://doi.org/10.1007/978-3-658-10902-8

50. Chen, C. (2001). An investigation of noise and vibration in an automotive power steering system. Ph.D. Dissertation. The Ohio State University, Columbus, Ohio.

51. Baum, H., Erzberger, B. (2015). Pulsationsdämpfer für ein hybrides Pumpensimulationsmodell. VDI-Berichte, 195–204. (in German).

52. Hoffmann, D. (1976). Die Dämpfung von Flüssigkeitsschwingungen in Ölhydrauli-kleitungen, VDI-Verlag, Düsseldorf. (In German).

53. Beynart, V. (1999). Pulsation dampening in suction and discharge systems for PD pumps. *World Pumps*, 20–24.

54. Koegler, A., Haselmann, D., Alt, N., & Schluecker, E. (2016). Experimental characteri-zation of a flow-through pulsation damper regarding pressure pulsations and vibra-tions. *Chemical Engineering & Technology, 40*(1), 162–169. https://doi.org/10.1002/ceat.201600175

55. Chai, L., Jiao, Z., Xu, Y., Zheng, H. (2016). A compact design of pulsation attenuator for hydraulic pumps. In *Proceedings of IEEE International Conference on Aircraft Utility Systems (AUS)* (pp. 1111–1116). https://doi.org/10.1109/AUS.2016.7748225

56. Bach, D., Masselter, T., & Speck, T. (2017). Damping of pressure pulsations in mobile hydraulic applications by the use of closed cell cellular rubbers integrated into a vane pump. *Journal of Bionic Engineering, 14*(4), 791–803. https://doi.org/10.1016/S1672-6529(16)60444-4

57. Esser, J. (1995). Pulsationsdämpfer für die Mobilhydraulik. O+P Ölhydraulik und Pneumatik (pp. 824–827). (in German).

58. Mikota, J. (2001). A novel, compact pulsation compensator to reduce pressure pulsa-tions in hydraulic systems. *World Scientific, 45*, 69–78.

59. Kitajima, D., Machimura, H., Munakata, A., Nemoto, M., Yamauchi, H. (2013). Fluid pressure pulsation damper mechanism and high-pressure fuel pump equipped with fluid pressure pulsation damper mechanism. US patent US,83,66,421,B2.

60. Paidoussis, M., & Li, G. (1993). Pipes conveying fluid: A model dynamical problem. *Journal of Fluids and Structure, 7*(2), 137–204. https://doi.org/10.1006/jfls.1993.1011

61. Wang, L., & Ni, Q. (2006). A note on the stability and chaotic motions of a restrained pipe conveying fluid. *Journal of Sound and Vibration, 296*(4–5), 1079–1083. https://doi.org/10.1016/j.jsv.2006.03.016

62. Paidoussis, M. (2003). *Fluid-structure interactions: Slender structures and axial flow* (Vol. 2, 942 p.). Elsevier Academic Press. ISBN 9780123973337.

63. Fornarelli, F., Lippolis, A., Oresta, P., & Posa, A. (2017). A computational model of axi-al piston swashplate pumps. *Energy Procedia, 126*, 1147–1154. https://doi.org/10.1016/j.egypro.2017.08.314

64. Wang, L. (2008). A further study on the non-linear dynamics of simply supported pipes convey-ing pulsation fluid. *International Journal of Non-Linear Mechanics, 44*(1), 115–121. https://doi.org/10.1016/j.ijnonlinmec.2008.08.010

65. Chen, C. (2012). Noise and vibration in complex hydraulic tubing systems. In *Continuum mechanics—Progress in fundamentals and engineering applications* (pp. 89–105). ISBN 978-953-51-0447-6.

66. Gorman, D., Reese, J., & Zhang, Y. (2000). Vibration of a flexible pipe conveying vis-cous pul-sating fluid flow. *Journal of Sound and Vibration, 230*(2), 379–392. https://doi.org/10.1006/jsvi.1999.2607

67. Luczko, J., & Czerwiński, A. (2014). Parametric vibrations of pipes induced by pulsat-ing flows in hydraulic systems. *Journal of Theoretical and Applied Mechanics, 52*(3), 719–730.

68. Dai, H., Wang, L., Qian, Q., & Ni, Q. (2013). Vortex-induced vibrations of pipes con-veying fluid in the subcritical and supercritical regimes. *Journal of Fluids and Structures, 39*, 322–334. https://doi.org/10.1016/j.jfluidstructs.2013.02.015

69. Lin, Z. (1986). *Engineering computation of deformation force under forging* (pp. 20–40). Me-chanical Industry Press.

70. Sewall, J., Wineman, D., Herr, R. (1973). An investigation of hydraulic-line reso-nance and its attenuation. National Aeronautics and Space Adninistration (NASA), NASA TM X-2787 (78 p.).

71. Van den Horn, B., & Kuipers, M. (1988). Strength and stiffness of a reinforced flexible hose. *Applied Scientific Research, 45*(3), 251–328. https://doi.org/10.1007/BF00384690

72. Czerwiński, A., & Łuczko, J. (2015). Parametric vibrations of flexible hoses excited by a pulsating fluid flow. *Part II: Experimental research, Journal of Fluids and Struc-tures, 55,* 174–190. https://doi.org/10.1016/j.jfluidstructs.2015.03.007

73. Xu, Y., Johnston, D., Jiao, Z., & Plummer, A. (2014). Frequency modelling and solution of fluid-structure interaction in complex pipelines. *Journal of Sound and Vibration, 333*(10), 2800–2822. https://doi.org/10.1016/j.jsv.2013.12.023

Sources of Pressure Pulsation in the Machines Hydraulic Systems

2

This chapter will specifically analyze the sources of pressure pulsations in an operating hydrostatic system. The main causes of pressure pulsations in the hydrostatic system are:

- Pulsation of displacement pump capacity;
- Transient states of the hydrostatic system (such as the start-up of the hydrostatic transmission);
- The external mechanical vibration's action on the elements of the hydraulic system, including hydraulic valves;
- Wave phenomena occurring under certain conditions in hydraulic lines, also called long hydraulic lines.

In order to better identify the components of the amplitude-and-frequency spectrum of pressure pulsation in the hydrostatic system, it is helpful to analytically describe the pulsation capacity of the displacement pump with which the hydrostatic system is fitted. An operating positive displacement pump generates a pulsatile flow which is recorded as a pressure pulsation in the hydrostatic system. The frequencies of this pulsation depend on the number of pump displacement elements (number of pistons, vanes, and teeth) and the speed of the pump drive shaft.

© The Author(s), under exclusive license to Springer Nature Switzerland AG 2024   31
M. Stosiak and M. Karpenko, *Dynamics of Machines and Hydraulic Systems*, Synthesis Lectures on Mechanical Engineering, https://doi.org/10.1007/978-3-031-55525-1_2

2.1   Pressure Pulsation Caused by the Pulsation of the Displacement Pump Capacity

Pressure pulsation is a significant phenomenon that emerges as a result of the dynamic operation of displacement pumps. These pumps, known for their ability to move fluids by changing the volume of a fluid chamber, inherently introduce pulsations in the fluid flow. This sub-chapter delves into an in-depth exploration of the pressure pulsations stemming from the oscillating displacement pump capacity. By understanding the mechanisms driving these pulsations, engineers and researchers can devise effective mitigation strategies and optimize pump designs for various applications.

The kinematics of a positive displacement pump generates pulsatile flow by the pump, which creates pressure pulsations in the hydraulic system. In our research and other literature [1–5], it is stated that the pressure pulsation acts on the components of the hydrostatic system, inducing their vibrations and contributing to the increased noise of the entire system. The dimensionless coefficient of capacity irregularity, also called the pulsation coefficient, can be used to describe the pulsation of displacement pump capacity [6]:

$$\delta = \frac{Q_{max} - Q_{min}}{Q_{avg}},$$
(2.1)

where Q_{max} and Q_{min}—the momentary maximum and minimum capacity, respectively, and Q_{avg}—the average pump capacity.

Figure 2.1 shows the coefficient δ as a function of the number of pump displacement elements for equal design variations.

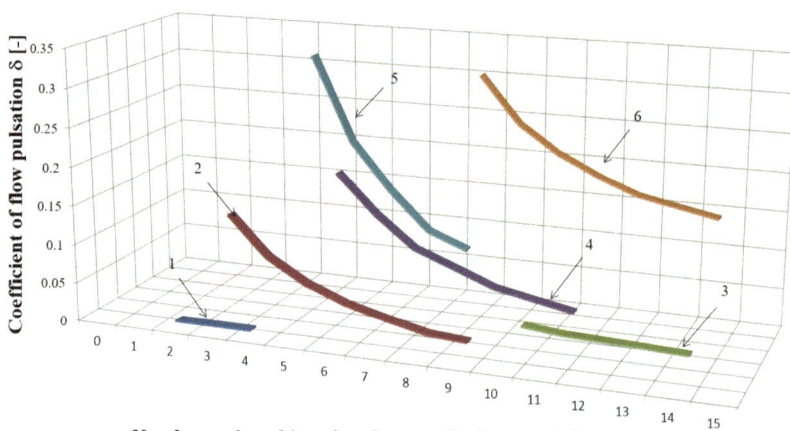

Fig. 2.1 Capacity irregularity coefficient of various design variants of positive displacement pumps: 1—screw; 2—piston with an odd number of pistons; 3—gear with internal meshing; 4—vane; 5—piston with even number of pistons; 6—gear with external meshing

As shown in Fig. 2.1, the capacity irregularity coefficient is typically the highest for multi-piston pumps with an even number of pistons. They outperform even external-meshing gear pumps in terms of capacity pulsation. The use of multi-piston pumps with an odd number of pistons results in a reduction of pump capacity pulsations. The lowest capacity irregularity value is characteristic of screw pumps, however, due to relatively low operating pressures, they are rarely used in hydrostatic systems. The utilization of the irregularity coefficient for the analysis of the capacity pulsation as a function of time or frequency is considered inconvenient.

Using the representation of the capacity pulsation in the time domain and its transformation in the frequency domain, more useful information can be obtained to create a comprehensive identification of the capacity pulsation.

Capacity pulsations are described by various mathematical models with forms depending on the design variation of the positive displacement pump. In the mechanical engineering industry as well as mobile hydraulics, it is common to observe positive displacement pumps: vane, gear, and multi-piston pumps. The literature shows analytical models describing the pulsation of capacity (instantaneous capacity) of displacement pumps in the time domain (or the angle of rotation of the pump drive shaft), which allows it to be presented in the form of an amplitude-and-frequency spectrum convenient for analysis. This spectrum contributes to the analysis of the amplitude-and-frequency spectrum of pressure pulsation caused by displacement pump capacity pulsation.

2.1.1 Single-Acting Vane Pump Capacity Pulsation

The form of the analytical model of the instantaneous capacity of a single-acting vane pump, taking into account the thickness of the blades, can be expressed by the following equation [6]:

$$Q = \left[r_l - \frac{e}{2} \cdot (1 + \cos\phi_l) \right] \cdot \omega_l \cdot b_l \cdot e \cdot (1 - \cos\phi_l) - e \cdot \omega_l \cdot b_l \cdot s_l \times$$
$$\times \left\{ \sin\phi_l + \sin(\phi_l + \alpha_l) + \ldots + \sin\left[\phi_l + \left(\frac{z_l}{2} - 1 \right) \cdot \alpha_l \right] \right\}, \tag{2.2}$$

where, r_l—stator bore radius, s_l—vane thickness, b_l—vane width, z_l—number of vanes, α_l—center angle of vane spacing in the rotor, ω_l—angular velocity of the pump shaft, ϕ_l—rotor rotation angle, e—eccentricity.

The analytical model of displacement pump pulsation can be implemented using specialized software. Many special-purpose programs have built in libraries for hydraulic components, including pumps. However, these models exclude capacity pulsations and only provide an average value of the capacity generated by the pump. An exact representation of the instantaneous capacity of the pump may be made in the Simulink package in the Matlab environment. In order to perform a computer simulation of the instantaneous capacity of the vane pump, the technical documentation of the V3-63 pump manufactured by Fabryka Elementów Hydrauliki Ponar-Wadowice SA was used. The block diagram of the Matlab-Simulink program represented by Eq. (2.2) is shown in Fig. 2.2. This allows the procurement of a time waveform of the pump capacity for the design parameters entered by the user (i.e., stator bore radius, vane width and thickness, and number of vanes) and operational (i.e., speed at the drive shaft, the eccentricity) and recorded in the working space for further analysis.

Based on Eq. (2.2), the simulation (according to the diagram in Fig. 2.2) waveform of the instantaneous capacity of the V3-63 vane pump is obtained, as shown in Fig. 2.3a. In terms of frequency analysis, the amplitude-and-frequency spectrum of the time waveform is important. This can be achieved using a description with the Fourier series and a specialized approach. The amplitude-and-frequency spectrum of the V3-63 pump capacity pulsations obtained with the Origin program is shown in Fig. 2.3b.

The values of the frequencies of the individual harmonics of the spectrum are described by the following relationship:

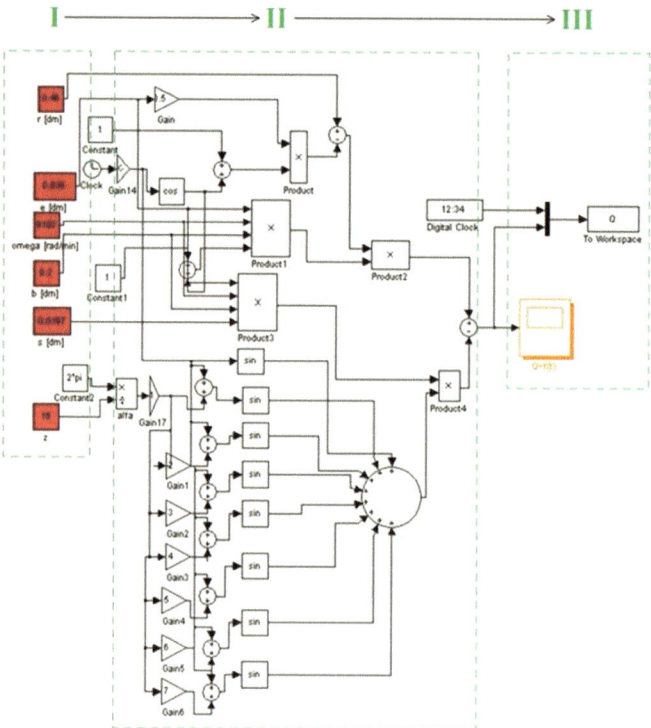

Fig. 2.2 Block diagram of Matlab-Simulink representing the theoretical model of V3-63 vane pump capacity pulsation; area I—entering data, area II—performing calculations according to Eq. (2.2), area III—calculation results in the form of a time waveform and exporting results to the workspace

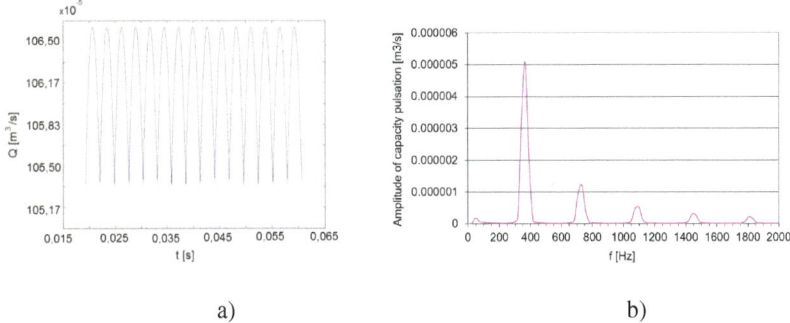

a)   b)

Fig. 2.3   The charts regarding V3-63 vane pump: **a** theoretical waveform of instantaneous capacity as a function of time for the V3-63 vane pump (1450 rpm); **b** the amplitude-and-frequency spectrum of the instantaneous capacity of the V3-63 pump

$$f_K = \frac{n_p \cdot z_l \cdot K}{60} \cdot HZ, \tag{2.3}$$

where n_p—rotational speed of the pump shaft [rpm], K—consecutive component number $= 1 \ldots K_i$.

The pump under consideration has 15 blades, and the rotational speed of the pump shaft is 1450 rpm. As shown in Fig. 2.4, the first component of the spectrum ($K = 1$) occurs at frequency $f_1 = 362$ Hz. The subsequent components are its multiples.

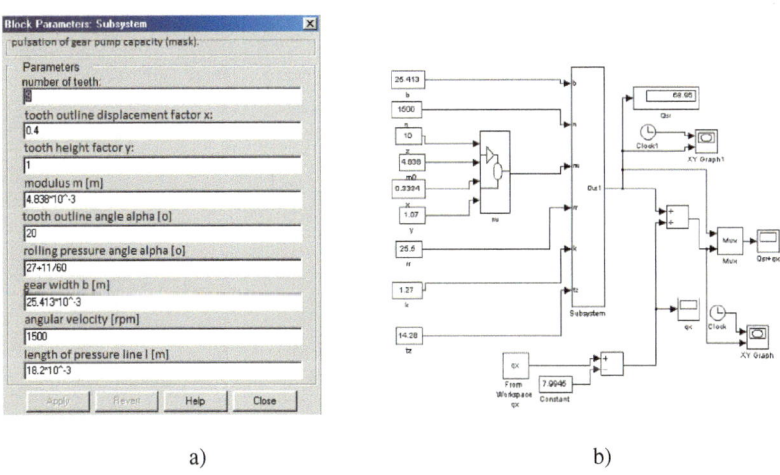

a)   b)

Fig. 2.4   Simulation details for gear pump: **a** view of the interface for entering construction and operational data of the gear pump; **b** a block model of the capacity pulsation of the 2110 gear pump, presented using the Simulink package

2.1.2 Capacity Pulsation of an External-Meshing Gear Pump

Based on the source materials [6], the instantaneous capacity of an external-mesh gear pump can be represented as:

$$Q = \frac{b_z \cdot \omega_l}{2} \cdot \left[r_{w1}^2 + \frac{r_{t1}}{r_{t2}} \cdot r_{w2}^2 - r_{t1} \cdot (r_{t1} + r_{t2}) - \left(1 + \frac{r_{t1}}{r_{t2}} \right) \cdot u^2 \right] \qquad (2.4)$$

where b_z—width of the wheels, ω_l—angular velocity of the gear wheel, r_w—radius of the top wheel, r_t—radius of the rolling wheel, $u = r_z \cdot \varphi_z$—instantaneous distance of the contact point of the teeth, moving along the buttress line, starting from the mesh pole, r_z—radius of the main circle, φ_z—angle of rotation of the gears.

The theoretical model reflected by Eq. (2.4) has been implemented in Matlab's Simulink package. For a computer simulation of the instantaneous capacity of the gear pump, the technical documentation of the 2110 pump manufactured by WZMB im. Waryński is used. The diagram of a specially developed interface that allows the user to enter the necessary design and operational data and the Matlab-Simulink blocks representing Eq. (2.4) is presented in Fig. 2.4.

Based on Eq. (2.4) and the simulation (according to the diagram in Fig. 2.4b), the waveform of the instantaneous capacity of the gear pump 2110 is obtained, as shown in Fig. 2.5a. To perform a discrete analysis of the amplitude-and-frequency spectrum, the Origin program is used. The amplitude-and-frequency spectrum of the capacity pulsation of pump 2110 is shown in Fig. 2.5b.

The dominant frequency, according to Eq. (2.3), is $f_1 = 250$ Hz because the number of teeth in the pump 2110 is $z_z = 10$, and the rotational speed of the pump shaft is $n_p = 1500$ rpm.

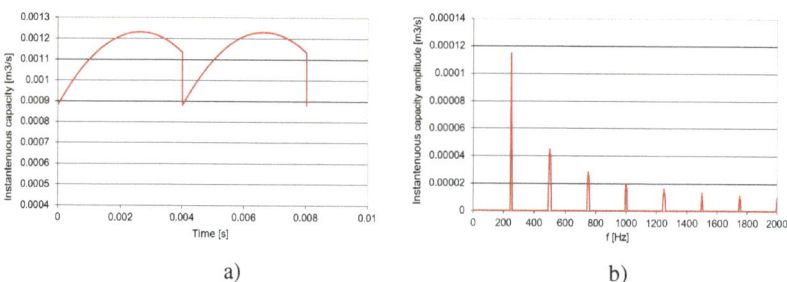

a)   b)

Fig. 2.5 The charts regarding gear pump 2110: **a** theoretical instantaneous capacity as a function of time for pump 2110 (1500 rpm); **b** the amplitude-and-frequency spectrum of the instantaneous capacity of the gear pump 2110

2.1.3 Capacity Pulsation of a Multi-piston Pump

The instantaneous capacity of an axial piston pump with a swinging impeller is determined by a theoretical relationship [6], for i-pistons being simultaneously in the discharge zone:

$$Q = \omega_p \cdot A_t \cdot R_t \cdot \sin\alpha_w \cdot \sum_{K=0}^{i-1} \sin\left(\varphi_t + K \cdot \frac{2 \cdot \pi}{z_t}\right) \tag{2.5}$$

where ω_p—the angular velocity of the pump shaft, A_t—the cross-sectional area of the pump's displacement chamber, R_t—the radius of spacing of the pistons in the pump, α_w—the angle of deflection of the pump rotor, φ_t—the angle of rotation of the piston pump shaft. In addition, i depends on whether the number of pistons z_t is even or odd.

For an odd number of pistons, as in the case of the PNZ-25 pump with seven pistons, $i = (z_t + 1)/2$ if $0 \le \varphi_t \le \frac{\alpha_t}{2}$ or $i = (z_t - 1)/2$ for $\frac{\alpha_t}{2} \le \varphi_t \le \alpha_t$, where α_t is the angular pitch $\alpha_t = 2\pi/z_t$. Thus, for a pump shaft angle φ_t between 0 and $\alpha_t/2$ i takes the value 4, while when the angle φ_t is between $\alpha_t/2$ and α_t, $i = 3$.

Taking this into account, Eq. (2.5) can be written:

$$Q = \omega_p \cdot A_t \cdot R_t \cdot \sin\alpha_w \cdot (\sin\varphi_t + \sin(\varphi_t + \alpha_t) + \sin(\varphi_t + 2\alpha_t) + \sin(\varphi_t + 3\alpha_t)) \tag{2.6}$$

where

$$0 \le \varphi_t \le \frac{\alpha_t}{2} \tag{2.7}$$

and

$$Q = \omega_p \cdot A_t \cdot R_t \cdot \sin\alpha_w \cdot (\sin\varphi_t + \sin(\varphi_t + \alpha_t) + \sin(\varphi_t + 2\alpha_t)) \tag{2.8}$$

where

$$\frac{\alpha_t}{2} \le \varphi_t \le \alpha_t. \tag{2.9}$$

Equations (2.6) and (2.8) describe the instantaneous capacity of the pump under consideration, assuming that there are four and then three pistons, respectively, alternating in the discharge phase.

The theoretical model reflecting Eqs. (2.6 and 2.8) was implemented, in Matlab 6.5 using the Simulink package. In order to carry out a computer simulation of the instantaneous capacity of a multi-piston pump, the technical documentation of the PNZ-25 pump manufactured by Fabryka Maszyn Budowlanych Bumar-Hydroma S.A. Szczecin. As a result, the time course of the instantaneous capacity of the PNZ-25 pump was obtained, shown in Fig. 2.6a. To analyse the components of the time waveform, it is convenient

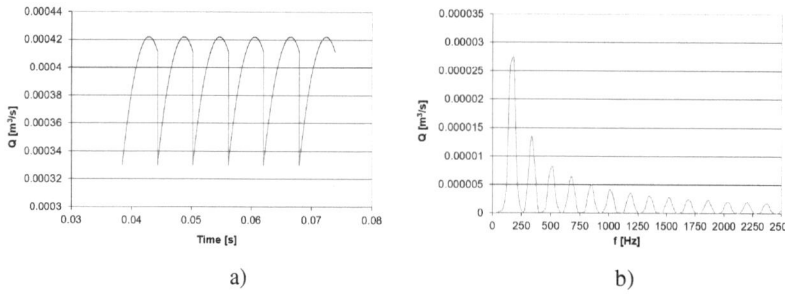

Fig. 2.6 The charts regarding PNZ-25 pump: **a** theoretical course of instantaneous capacity as a function of time for a PNZ-25 multi-piston pump (1500 rpm); **b** amplitude-frequency spectrum of the instantaneous output of the PNZ-25 pump

to use an amplitude-frequency spectrum. Figure 2.6b shows the performance pulsation spectrum of the PNZ-25 pump limited to a frequency of 2500 Hz.

The dominant frequency in this case is 175 Hz, as the number of pistons in the PNZ-25 pump is $z_t = 7$ and the pump shaft speed $n_p = 1500$ rpm.

2.1.4 Comparison of Pumps Capacity Pulsation

As shown in Figs. 2.3b, 2.5b and 2.6b, the first harmonic of the spectrum of instantaneous capacity ranges from 175 to 362 Hz, depending on the rotational speed of the pump shaft and the number of displacement elements. In each case, the first component is determined by Eq. (2.3) being dominant over the entire spectrum. According to Fourier's theorem, each of the functions graphically presented in Figs. 2.3a, 2.5a and 2.6a can be expanded into a Fourier series, and the values of the successive components have values corresponding to the successive amplitudes in the spectra shown in Figs. 2.3b, 2.5b and 2.6b.

Comparison of the amplitude-and-frequency spectra of the considered pumps (Fig. 2.7), shows that the highest value of the first component of capacity pulsation is characteristic of the gear pump 2110, and the lowest value of the vane pump V3-63.

As shown in Fig. 2.7, the comparison of coefficients δ indicates that the non-uniformity of the vane pump capacity can be greater than the capacity pulsation of the piston pump. As reported Kudźma [7], the actual capacity pulsation of the vane pump may be ten times greater than that resulting from the above-mentioned theoretical analysis (Fig. 2.3a). The reasons were described in the publication Kudźma [7], regarding the vane vibration, which was an additional source of pulsation of the capacity of the vane pump.

To examine the share of the individual components of the capacity pulsation spectrum in the average capacity, the non-uniformity coefficient of capacity can be related to the individual components [7]:

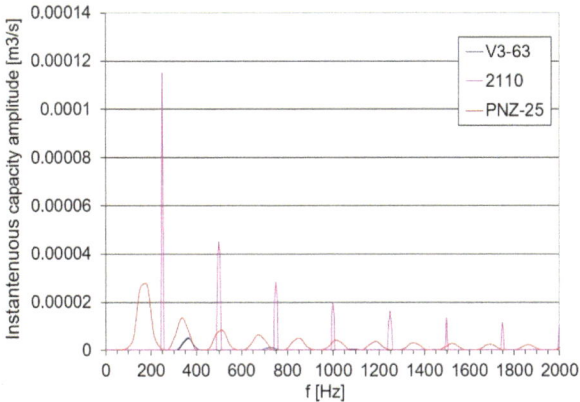

Fig. 2.7 Comparison of the amplitude-and-frequency spectrum of the instantaneous capacity of positive displacement pumps: V3-63—vane pump with $z_l = 15$ vanes and unit capacity $q = 0.000046$ m^3/rev, 2110—gear pump with $z_z = 10$ teeth and $q = 0.000045$ m^3/rev, PNZ-25—multi-piston axial pump with $z_t = 7$ pistons and $q = 0.000023$ m^3/rev

$$\delta_K = \frac{q_K}{Q_{\text{avg}}} \cdot 100\% \tag{2.10}$$

Figure 2.8 shows a cumulative graph for the first three components of the amplitude-and-frequency spectrum of the instantaneous capacity of the basic types of positive displacement pumps: single-acting V3-63 vane, 2110 gear, and PNZ-25 multi-piston.

Capacity pulsation is the cause of pressure pulsation, which can indicate an increase in the noise of hydraulic systems, non-uniform movement of the hydraulic receiver [8, 9], excitation of vibrations of hydraulic lines, and other elements of the hydraulic system. These spectral analyzes show that the basic component of the capacity pulsation spectrum, as a result of the kinematics of the movement of the pump's displacement elements, and due to the pressure, which is in the range from 175 Hz to approx. 350 Hz. Also, it

Fig. 2.8 The amplitude-and-frequency spectrum of instantaneous capacity related to the theoretical capacity of displacement pumps for the first three harmonic components

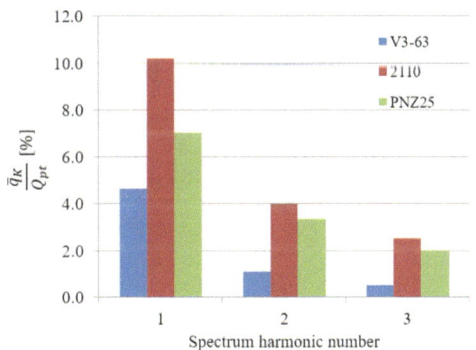

depends on the number of displacement elements and the speed of the pump drive shaft. In addition, there are other sources of pressure pulsations in the hydraulic system, the frequency of which differs from those presented above.

2.2   Pressure Pulsation Caused by Transients of Hydraulic Systems

An inherent phase of operation of each hydraulic system is the transition phase, in which pressure pulsations also occur. This phase occurs, e.g., during start-up, the shutdown of the system, sudden change of external load, sudden change of the speed of the hydraulic receiver, or direction of its movement. Transient states are accompanied by, among others, increased dynamic load of elements, dynamic pressure excess, and increased noise, as well as mechanical vibrations of hydraulic system elements.

The problem of describing systems in these states is widely discussed in the literature [10–15]. A comprehensive approach to the dynamic states of hydraulic systems and components is presented in ref. Yang et al. [15], Khalil [16], Totten and Negri [17], where the authors describe in detail various models for the analysis and synthesis of the dynamics of hydraulic systems. Dynamic quality indices are used as criteria for assessing dynamic states [15]. Among these criteria, the following are commonly used in addition to integral criteria: dynamic excess coefficient (overshoot coefficient), reduced damping coefficient (which is a measure of energy dispersion), response time, transient process time, natural frequency.

During a transient process, e.g., the start-up of a hydrostatic transmission, pressure pulsations also appear in the low-frequency range. The frequency of pressure pulsations depends on how the start-up process takes place, i.e., whether the maximum valve is or is not involved [18, 19]. If the start-up process of the hydrostatic transmission (as a response to a surge forcing caused by the flow rate) is carried out without the participation of the maximum valve, the frequency of the pressure pulsation depends, among others, on the capacitance of the system, the active area of the cylinder, the mass reduced per piston, and the reduced damping coefficient, which is a function of, e.g., leakage coefficient, coefficient of friction, cylinder active area, capacitance and mass reduced per cylinder piston. The time waveform of the pressure, and therefore its pulsation, depends significantly on the reduced damping coefficient. If the reduced damping coefficient is ≤1, the start-up process is oscillatory, that is damped vibration with overshoot and rapid pressure increase. However, if the reduced damping coefficient is >1, then the start-up process is asymptotic, with pressure slowly increasing to p_u in a steady-state motion, resulting mainly from the load on the external hydraulic receiver.

However, during the start-up of the hydrostatic transmission involving a maximum valve, this valve's control element may be forced to vibrate (e.g., a poppet), and, consequently, additional harmonic components may occur in the pressure pulsation spectrum.

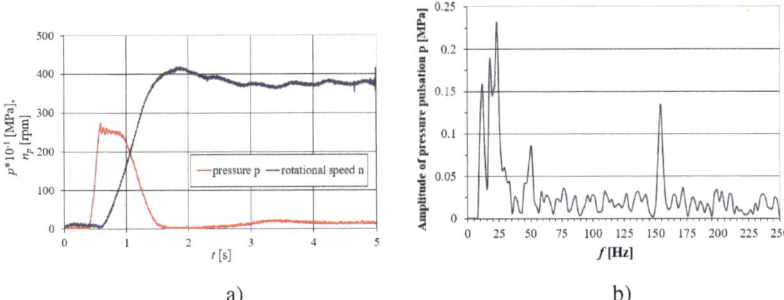

a)   b)

Fig. 2.9 The charts regarding fluid pressure pulsation: **a** the course of pressure and rotational speed of the slewing mechanism of a crane with a hydrostatic transmission during start-up with the use of the maximum valve; **b** the amplitude-and-frequency spectrum of pressure pulsations during the hydrostatic transmission start-up process involving the maximum valve [20]

Therefore, it is important to determine the natural frequency of the control element of the maximum valve. Ref. Kudźma [7], provides details of the frequency of undamped free vibrations of a single-stage maximum valve as a function of valve design and operating parameters. These included valve reinforcement, flow coefficient, valve seat diameter, the angle between the poppet surface and the axis of symmetry, working medium density, pressure work, and reduced mass of the poppet and spring. In the case of the impact of the hydraulic system, it considers the impact of the capacitance of the lines and hydraulic oil, and the concept of reduced stiffness, which includes not only the capacitance but also the stiffness of the valve spring. The results of the frequency of undamped natural vibrations of the valve head are consistent with the experimental results. Figure 2.9a shows an example of a time waveform of the pressure and the increase in rotational speed of the hydraulic motor. Figure 2.9b shows the amplitude-and-frequency spectrum of pressure pulsations during the start-up of the hydrostatic transmission of the truck crane slewing mechanism with a maximum valve.

The presented waveforms show that the dominant frequency corresponds to the natural frequency of the relief valve, i.e., at approx. 25 Hz lying in the range of 5–25 Hz and to the resonance of the valve and the natural frequency of the crane slewing mechanism. This range of valve resonance frequencies has been confirmed in the literature Backé [21], where experimental testing and theoretical analysis of the pressure variability and the position of the poppet in the single-stage relief valve have been conducted. By recording the instantaneous course of the pressure and the position of the relief valve poppet in a hydraulic system with a surge forcing of the flow, it can be concluded that the natural frequency of such a valve is approx. 20 Hz. At the same time, the report noted that a change in this frequency is obtained by changing the active conductivity of the damping gap, determined by the gap's dimensions. An increase in the active conductivity of the damping gap leads to an increase in the natural frequency of the valves, and at the same time to the valve falling into vibrations—to the loss of valve stability. In addition, the spectrum in Fig. 2.9b shows a component of approx. 169 Hz, which stems from the displacement pump capacity pulsation (in the tested gear, an axial multi-piston pump was used for which the number of pistons $z = 7$, and the speed on the drive shaft was $n = 1450$ rpm).

2.3 Pressure Pulsation Caused by External Mechanical Vibrations

A working machine fitted with hydraulic components and systems during operation is a source of mechanical vibrations with a wide frequency range. These excitations, in the form of mechanical vibrations, are transferred to the environment and often cause adverse effects on humans (operator) and the hydraulic system located in that environment [20, 22–24]. Therefore, the components of the hydraulic system of the working machine, such as pumps, valves, lines, and receivers, are subjected to mechanical vibrations. This may cause additional disturbances, which are evident by the appearance of changes in the spectrum of pressure pulsations [23, 25] and in the operation of hydraulic systems, which may lead to reduced accuracy of the actuator positioning, uneven operation, shorten the machine's life, occasionally increased emitted noise level, etc. In terms of the correct operation of hydraulic valves, external mechanical vibrations in the resonance range of the valve control element (e.g., spool, poppet, ball) and accordance with the direction of movement of this element.

The resonant frequencies of the controls of typical single-stage hydraulic valves are usually below 100 Hz [2, 26], while the resonance frequencies of the controls of single-stage micro-valves are much higher, ranging from approx. 600 Hz to approx. 900 Hz [27]. These may coincide with vibrations generated by machines or devices equipped with hydraulic systems. The setting elements of hydraulic valves can then be forced to vibrate, which may cause periodic changes in the size of the valve's throttling gaps. Consequently, the appearance of changes in the pressure pulsation spectrum occurs in the form of the spectrum harmonics corresponding to the frequencies of the vibrating element (e.g., spool, poppet, ball) [27]. Figure 2.10a highlights this problem, which is extensively discussed in the following chapters. Additionaly at Fig. 2.10b shows an example of an amplitude-and-frequency spectrum of pressure pulsation in a system, in which external mechanical vibrations affect the conventional single-stage relief valve with the symbol DBDH 6 G18/100.

As shown in Fig. 2.10a, a clear relationship is observed between the pressure pulsation and the external mechanical vibrations. In the amplitude-and-frequency spectrum of pressure pulsations in the hydraulic microsystem in which the relief microvalve is subjected to external mechanical vibrations with a frequency of 600 Hz (black line), there is

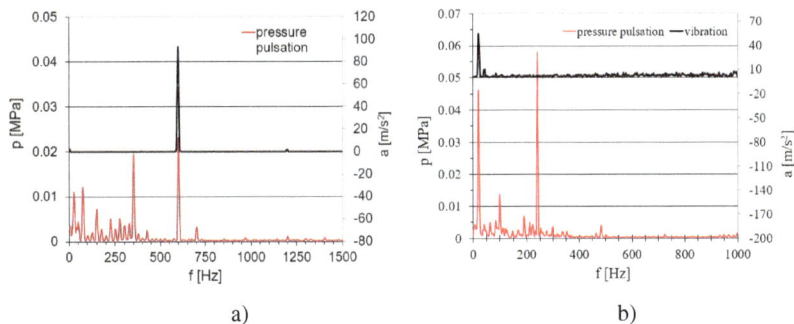

Fig. 2.10 The amplitude-and-frequency spectrum of external mechanical vibrations and pressure pulsations (mean pressure $p_{avg} = 5$ MPa, the flow rate through the valve $Q = 1.083 \times 10^{-4}$ m³/s (6.5 dm³/min)): **a** frequency of external mechanical vibrations $f = 600$ Hz; **b** frequency of external mechanical vibrations $f = 20$ Hz

a component corresponding to this frequency (red line). The spectrum of pressure pulsations is dominated by the 600 Hz component. Other components are also present, among which is the distinctive frequency of approx. 350 Hz, and stems from the pulsation of the capacity of the geared micro-pump used. As mentioned earlier, this pulsation depends on the number of teeth (in this case, 14) and the rotational speed on the micro-pump's shaft (in this case, 1500 rpm). The vibration spectrum of a machine equipped with hydraulic components is wide enough to include the lower frequencies to which conventional valves may be sensitive.

Additionally, Fig. 2.10b shows the co-occurrence of pressure pulsation components corresponding to the frequency of external mechanical vibrations. In the amplitude-and-frequency spectrum of pressure pulsations in the system where the single-stage conventional relief valve affects external mechanical vibrations, there is a harmonic component with a frequency corresponding to these vibrations, $f = 20$ Hz. The dominant component of the spectrum is at approx. 250 Hz, and occurs due to the capacity of the positive displacement pump (gear pump with 10 teeth and rotational speed of the drive shaft of 1450 rpm).

2.4   Pressure Pulsation Caused by Resonance Phenomena in a Hydraulic Transmission Line (Pipeline)

Due to capacity fluctuations created by positive displacement pumps and changes in the geometrical dimensions of the flow gaps caused by the vibrating movement of the hydraulic valve control elements in the hydraulic system lines, pulsatile flows occur, namely, periodic fluctuations in the flow velocity and pressure of the working liquid. A pulsatile flow is a flow during which the flow velocity is the sum of two components:

the time-averaged component of the flow velocity and the variable component (usually harmonically or polyharmonically variable) [28]. However, the oscillatory (oscillating) flow is a special case of pulsatile flow, where the time-averaged component of the flow velocity is zero [28]. In a study of hydraulic systems in quasi-steady states, special attention is paid when the length of the hydraulic line is equal to or greater than the length of the pressure wave propagated therein [7], i.e., when dealing with a hydraulic transmission line (HTL).

A line can be treated as HTL when the following condition is met:

$$\lambda_f \leq \frac{v_0}{10 \cdot f_{max}}, \tag{2.11}$$

where: λ_f—wavelength, v_0—speed of pressure wave propagation in the line, f_{max}—maximum excitation frequency. In this case, the hydraulic line is treated as an element with distributed parameters, i.e., it is assumed that changes in pressure and flow rate propagate along the line axis with a finite speed in the form of traveling and reflected waves [13, 29, 30]. Due to certain conditions imposed on proportionally-controlled hydraulic systems, phenomena specific to systems with a HTL may sometimes occur in such systems.

These systems, which contain a proportional valve, require that the natural frequency f_o of the receiver is ≥ 3 Hz ($\omega_0 = 18.84$ s^{-1}). The natural frequency of a loaded hydraulic receiver depends on the volume of liquid contained between the valve and the receiver on the supply and discharge sides, the receiver's design parameters, and its load. In order to meet the conditions of the receiver's dynamic properties (the natural frequency value), the stick–slip phenomenon is eliminated, and the proportional valve should be installed as close as possible to the receiver and connected to the rigid lines. If the hydraulic power pack has a compact design with all elements located next to the pump, usually on the top plate of the tank, and the entire power pack is connected to the proportional valve and the receiver only uses supply and relief lines, then focus should be paid to the length of the supply line. This is due to the possibility of resonance phenomena that intensifies the pressure pulsation amplitudes. A diagram of a typical hydraulic system with a HTL is shown in Fig. 2.11.

There are two ways to describe systems with a HTL in quasi-steady or non-steady states. The first method assumes the harmonic excitation at the input end, this is waveform analysis based on frequency. The second method involves the analysis of transient processes as a function of time. The dynamic waveforms are analyzed in the frequency domain using the frequency method, which can be utilized to create the amplitude-frequency and phase-frequency characteristics of hydraulic systems with a HTL in a relatively simple way. In this method, the hydraulic line is characterized by a two-way element of the system with two inputs and two outputs: pressure $\tilde{p}$ and flow rate $\tilde{q}$, i.e., also called the hydraulic cross [7], Fig. 2.12.

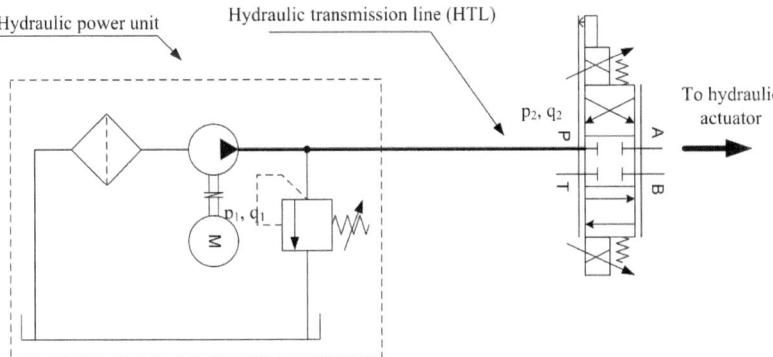

Fig. 2.11 A diagram of a hydraulic system with a HTL and a proportional valve: p_1 and q_1—deviations from the mean value of pressure and volumetric flow rate at the beginning of the HTL, respectively, p_2 and q_2—deviations from the mean value of pressure and volumetric flow rate, respectively, at the end of the HTL

Fig. 2.12 Hydraulic cross

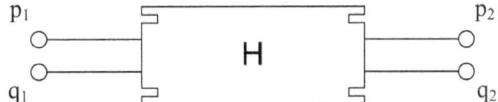

The individual symbols shown in Fig. 2.16 denote the following: Laplace transforms of pressure p_1 and flow rate q_1 at the beginning of the line, p_2 and q_2 Laplace transforms of pressure and flow rate at the end of the line, H—operator's transfer function.

In the quasi-steady state of the system, the instantaneous pressure and flow rate can be described as the sum of two components:

$$\tilde{p} = p_0 + p \text{ and } \tilde{q} = Q_0 + q \qquad (2.12)$$

where p_0 and Q_0—pressure and average flow rate over time, while p and q are deviations from the average value. The correct description of the pulsatile flow depends on the nature of the flow. In this case, the Reynolds number averaged over time contributes to the determination of the nature of the flow [28]. However, no relationship between the critical value of the Reynolds number, the frequency, and the amplitude of pulsations has been established so far. According to Ref. [28], turbulence can be predicted using the Reynolds number averaged over time. Carpinlioglu and Gundogdu [28] reported that the critical value of the Reynolds number $R_{ekr,p}$ for pulsatile flow defines the beginning of the so-called transition zone:

$$R_{ekr,p} = 2100 \qquad (2.13)$$

In contrast, the end of the interval of transient pulsatile flow, i.e., fully formed turbulence, can be adopted as in ref. Carpinlioglu and Gundogdu [28], when the value of the Reynolds number $R_{ekr,pr}$ exceeds:

$$R_{ekr,pr} = 8330 \tag{2.14}$$

In the frequency method, i.e., when a quasi-steady state under harmonic forcing is considered, the transmittance matrix takes the form of [7, 21]:

$$H = \begin{bmatrix} h_{11} & h_{12} \\ h_{21} & h_{22} \end{bmatrix} \tag{2.15}$$

The terms of the matrix H are given by the following relationships [7, 21]:

$$h_{11} = \cosh(T \cdot \Psi_z \cdot j\omega) \tag{2.16}$$

$$h_{12} = Z_c \cdot \Psi_z \cdot \sinh(T \cdot \Psi_z \cdot j\omega) \tag{2.17}$$

$$h_{21} = \frac{1}{Z_c \cdot \Psi_z} \sinh(T \cdot \Psi_z \cdot j\omega) \tag{2.18}$$

$$h_{22} = \cosh(T \cdot \Psi_z \cdot j\omega) \tag{2.19}$$

where the characteristic impedance of the line Z_c is expressed by the following relation:

$$Z_c = \frac{\rho_0 \cdot v_0}{\pi \cdot R^2} \tag{2.20}$$

where R—inner radius of the line, ω—excitation frequency, v_0—pressure wave propagation velocity, ρ_0—liquid density, j—imaginary unit and the time constant T depending on the parameters of the line, which is defined by the following equation:

$$T = \frac{L}{v_0} \tag{2.21}$$

where L—length of the hydraulic pipeline.

In the frequency-based modeling of a hydraulic long line, it is important to consider the influence of the viscosity of the working medium. For this purpose, the viscosity function is introduced Ψ_z, describing the friction resistance, and written in the form of the following relationship [7]:

$$\Psi_z = \frac{\Psi}{j \cdot \Omega} \tag{2.22}$$

$$\Psi = \varepsilon + j \cdot v_w \tag{2.23}$$

Dimensionless frequency Ω appears in Eq. (2.22) and is expressed by the relationship [7, 21]:

$$\Omega = \frac{\omega \cdot R^2}{v} \tag{2.24}$$

v—kinematic viscosity of the liquid.

In the description of pulsatile flows, the adequacy of the mathematical model's representation of reality depends on the adopted model of hydraulic resistance [7, 21]. The simplest model is one of a lossless line, in which the value of the viscosity function is assumed to be $\Psi_z = 1$. The most accurate model of hydraulic resistance for pulsatile flow (laminar and turbulent) is the variable resistance model that considers pressure losses as a function of frequency, also known as non-stationary friction losses. Often the correct results are obtained from a quasi-steady-state friction model. This can be used in systems with excitations of not very high frequencies (e.g., resulting from the pulsating capacity of displacement pumps or periodically changing geometrical dimensions of the flow slots of hydraulic valves, in which the control element vibrates at low frequencies; resonance frequencies of typical single-stage lift or diverter valves are usually up to 100 Hz). The appropriate viscosity function coefficients for this loss model are shown below [7]:

$$\varepsilon = 0.5 \cdot \Omega \cdot \sqrt{-1 + \sqrt{1 + \left(\frac{R_0}{\Omega}\right)^2}} \tag{2.25}$$

$$v_w = 0.5 \cdot \Omega \cdot \sqrt{1 + \sqrt{1 + \left(\frac{R_0}{\Omega}\right)^2}} \tag{2.26}$$

R_0—constant drag (resistance) determined with the Darcy-Weisbach formula:

$$R_0 = \frac{\lambda \cdot R_e \cdot \mu}{8 \cdot \pi \cdot R^4} \tag{2.27}$$

$\lambda = 64/R_e$—dimensionless coefficient of linear friction loss, R_e—Reynolds number, μ—dynamic viscosity of the liquid.

The HTL system can be represented by a complex one-port (Fig. 2.13).

Figure 2.13 shows that the element $H(j\omega)$ is treated as a two-port and described by Eqs. (2.15–2.19). Load impedance Z_k represents a one-port element of the system, which can

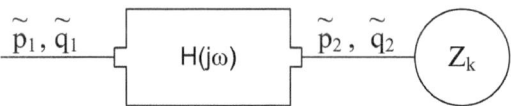

Fig. 2.13 Complex one-port for a system with a hydraulic long line

be, for example, a proportional valve treated in modeling as a constant drag (resistance). In general, hydraulic impedance is the relationship between harmonic changes in capacity and pressure and is a complex value that depends on the structure of the hydraulic system.

Regarding the block diagram and designations in Fig. 2.11, the matrix transmittance of the hydraulic long line can be presented as follows [7]:

$$\begin{bmatrix} p_1 \\ q_1 \end{bmatrix} = \begin{bmatrix} h_{11} \ h_{12} \\ h_{21} \ h_{22} \end{bmatrix} \cdot \begin{bmatrix} p_2 \\ q_2 \end{bmatrix} \tag{2.28}$$

When the load impedance is known, then it is defined as:

$$Z_k = \frac{p_2}{q_2} \tag{2.29}$$

The following transmittances of the system with a hydraulic long line are obtained

$$G_{p_1,q_1} = \frac{p_1}{q_1} = \frac{h_{11} Z_k + h_{12}}{h_{21} Z_k + h_{22}} \tag{2.30}$$

$$G_{p_2,p_1} = \frac{p_2}{p_1} = \frac{Z_k}{h_{11} Z_k + h_{12}} \tag{2.31}$$

$$G_{q_2,q_1} = \frac{q_2}{q_1} = \frac{1}{h_{21} Z_k + h_{22}} \tag{2.32}$$

$$G_{q_2,q_1} = \frac{q_2}{q_1} = \frac{1}{h_{21} Z_k + h_{22}} \tag{2.33}$$

The mathematical model of the system in Fig. 2.11 can be numerically solved after prior parameterization of the equation's coefficient values. This also allows for the graphical representation of the simulation solution, which is convenient for the interpretation of results for the obtained solution.

Model parameterization. $Q_0 = 8.33 \times 10^{-4}$ m^3/s—pump capacity (average over time), $p_0 = 50 \times 10^5$ Pa—pressure at the end of the hydraulic line (average over time), R = 0.009 m, $\rho_0 = 860$ kg/m^3, $v = 50 \times 10^{-6}$ m^2/s, $v_{o1} = 1288$ m/s (rigid line) according to [7], $v_{o2} = 800$ m/s (flexible line). The line ends with a proportional valve, which is treated as a throttling valve, hence, after linearization, the approximate final impedance is expressed as $Z_k = 2\Delta p_0/Q_0$, Δp_0—time average pressure difference through the valve, Q_0—time-averaged flow rate. In the case under consideration $Z_k = 1.2 \times 10^{10}$ N s/m^5.

Based on the system data and Eqs. (2.30–2.33), it is possible to determine a graphic of the transmittance waveform as a function of the length of the hydraulic line and the frequency of excitation (Figs. 2.14 and 2.15). The range of excitation frequencies in the presented graphs is adopted to include those caused by the kinematics of the operation of the pump displacement elements and those originating from the hydraulic valve control element excited to vibrations. On this basis, it is possible to determine what lengths of

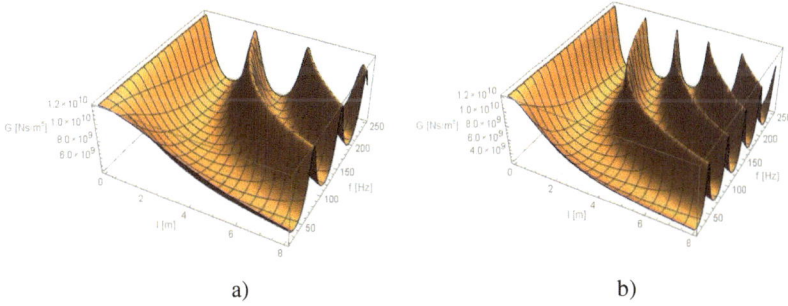

Fig. 2.14 Transmittance modulus G_{p2}, q_1 of the hydraulic system as a function of the length of the hydraulic line (rigid pipe) and the excitation frequency: **a** rigid pipe; **b** flexible pipe

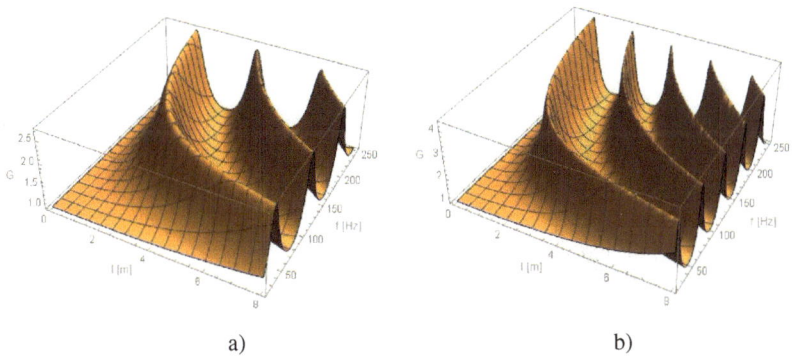

Fig. 2.15 Transmittance modulus G_{p2}, p_1 of the hydraulic system as a function of the length of the hydraulic line (rigid pipe) and the excitation frequency: **a** rigid pipe; **b** flexible pipe

hydraulic lines should be avoided to eliminate hydraulic resonance of pressure pulsations in the system.

For the selected excitation frequency, changes in the value of the transmittance modulus as a function of the length of the hydraulic line can be analyzed (Figs. 2.16 and 2.17).

In the next parts of the book, the attention was focused on the identification of various paths of transmission of internal hydraulic vibrations in pipelines and external mechanical vibrations to hydraulic valves and their controls. This will allow us to analyze the possibility of reducing the impact of vibrations on valves and hydraulic lines and their effect on changes in the pressure pulsation spectrum in the hydraulic system in line of energy-saving.

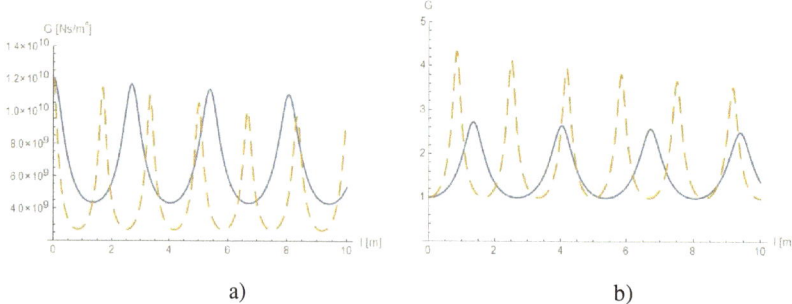

Fig. 2.16 Transmittance modulus as a function of pipe length L for flexible and rigid pipes. The excitation frequency $f_1 = 240$ Hz. (Solid line denotes a rigid pipe, dashed line denotes a flexible pipe): **a** G_{p2}, q_1; **b** G_{p2}, p_1

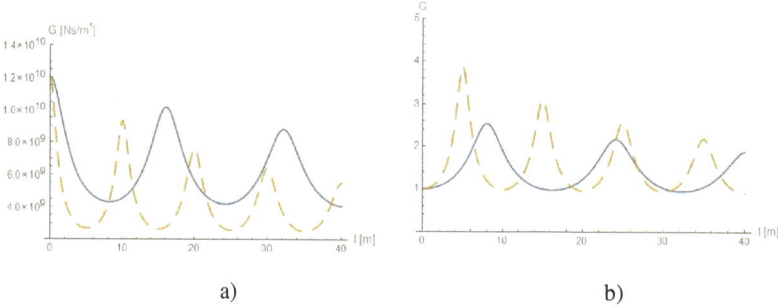

Fig. 2.17 Transmittance modulus as a function of pipe length L for flexible and rigid pipes. The excitation frequency of the first harmonic $f_1 = 40$ Hz. (Solid line denotes a rigid pipe, dashed line denotes a flexible pipe): **a** G_{p2}, q_1; **b** G_{p2}, p_1

References

1. Stosiak, M., Karpenko, M., Deptuła, A., Urbanowicz, K., Skačkauskas, P., Deptuła, A. M., Danilevičius, A., Šukevičius, Š., & Łapka, M. (2023a). Research of vibration effects on a hydraulic valve in the pressure pulsation spectrum analysis. *Journal of Marine Science and Engineering, 11*, 301. https://doi.org/10.3390/jmse11020301
2. Stosiak, M., Karpenko, M., Prentkovskis, O., Deptuła, A., Skačkauskas, P. (2023b). Research of vibrations effect on hydraulic valves in military vehicles. *Defence Technology*. ISSN 2214-9147. https://doi.org/10.1016/j.dt.2023.03.023
3. Kollek, W., Kudźma, Z., Rutański, J., & Stosiak, M. (2010). Acoustic problems relating to microhydraulic components and systems. *Archive of Mechanical Engineering, 57*(3). https://doi.org/10.2478/v10180-010-0016-9

4. Shi, L., Zhang, J., Yu, X., et al. (2023). Water hammer protection for diversion systems in front of pumps in long-distance water supply projects. *Water Science and Engineering, 16*(2), 211–218. ISSN 1674-2370. https://doi.org/10.1016/j.wse.2023.02.001

5. Zi, D., Wang, F., Wang, C., et al. (2021). Investigation on the air-core vortex in a vertical hydraulic intake system. *Renewable Energy, 177*, 1333–1345. ISSN 0960-1481. https://doi.org/10.1016/j.renene.2021.06.062

6. Stryczek, S. (2022). Napęd hydrostatyczny. WNT (314 p.). ISBN: 9788301193836 (In Polish).

7. Kudźma, Z. (2012). Tłumienie pulsacji ciśnienia i hałasu w układach hydraulicznych w stanach przejściowych i ustalonych. Oficyna Wydawnicza PWr (261 p.). (in Polish).

8. Pang, H., Wu, D., Deng, Y., et al. (2021). Effect of working medium on the noise and vibration characteristics of water hydraulic axial piston pump. *Applied Acoustics, 183*, 108277. ISSN 0003-682X. https://doi.org/10.1016/j.apacoust.2021.108277

9. Berestovitskiy, E., Ermilov, M., Kizilov, P., et al. (2015). Research of an influence of throttle element perforation on hydrodynamic noise in control valves of hydraulic systems. *Procedia Engineering, 106*, 284–295. ISSN 1877-7058. https://doi.org/10.1016/j.proeng.2015.06.037

10. Li, C.; Zhou, J.; Li, Y., et al. (2023). Dynamical modeling and physical analysis of pipe flow in hydraulic systems based on fractional variational theory. *Physics Letters A, 481*, 128999. ISSN 0375-9601. https://doi.org/10.1016/j.physleta.2023.128999

11. Munchhof, M. (2014). *Identification of dynamic systems* (705 p.). Springer. 9783642422676.

12. Yang, M., Yan, G., Zhang, Y., et al. (2023). Research on high efficiency and high dynamic optimal matching of the electro-hydraulic servo pump control system based on NSGA-II. *Heliyon, 9*(3), e13805. ISSN 2405-8440. https://doi.org/10.1016/j.heliyon.2023.e13805

13. Chiapponi, L., & Tanda, M. (2021). *Problems in hydraulics and fluid mechanics*. Springer Nature Switzerland AG. 3030513890.

14. Zhang, L., Wang, X., Wu, P., et al. (2023). Optimization of a centrifugal pump to improve hydraulic efficiency and reduce hydro-induced vibration. *Energy, 268*, 126677. ISSN 0360-5442. https://doi.org/10.1016/j.energy.2023.126677

15. Yang, Y., Qin, Z., & Zhang, Y. (2023). Random response analysis of hydraulic pipeline systems under fluid fluctuation and base motion. *Mechanical Systems and Signal Processing, 186*, 109905. ISSN 0888-3270. https://doi.org/10.1016/j.ymssp.2022.109905

16. Khalil, M. (2020). *Hydraulic systems volume 7: Modeling and simulation for application engineers*. COMPUDRAULIC LLC. 978-0997763430.

17. Totten, G., Negri, V. (2010). *Handbook of hydraulic fluid technology*. CRC. 9781420085266.

18. Liu, Y., Xu, Z., Hua, L., & Zhao, X. (2020). Analysis of energy characteristic and working performance of novel controllable hydraulic accumulator with simulation and experimental methods. *Energy Conversion and Management, 221*, 113196. ISSN 0196-8904. https://doi.org/10.1016/j.enconman.2020.113196

19. Bury, P., Stosiak, M., Urbanowicz, K., Kodura, A., Kubrak, M., & Malesińska, A. (2022). A case study of open- and closed-loop control of hydrostatic transmission with proportional valve start-up process. *Energies, 15*, 1860. https://doi.org/10.3390/en15051860

20. Kudźma, Z., & Stosiak, M. (2013). Reduction of infrasounds in machines with hydrostatic drive. *Acta of Bioengineering and Biomechanics, 15*(2), 51–64.

21. Backé, W. (1981). Schwingungserscheinungen bei Druckregelungen. Ölhydraulik und Pneumatik.

22. Zhang, J., Jiang, N., & Zhou, C. (2023). Quantitative evaluation method of human comfort under the influence of blast vibration based on human physiological indexes and its application. *Applied Acoustics, 202*, 109175. ISSN 0003-682X. https://doi.org/10.1016/j.apacoust.2022.109175

23. Sun, C., Liu, C., & Zheng, X. (2023). An analytical model of seated human body exposed to combined fore-aft, lateral, and vertical vibration verified with experimental modal analysis. *Mechanical Systems and Signal Processing, 200*, 110527. ISSN 0888-3270. https://doi.org/10.1016/j.ymssp.2023.110527
24. Sangbeom, W. (2022). Numerical and experimental analysis of vibroacoustic field of external gear pumps. Purdue University Graduate School. Thesis. https://doi.org/10.25394/PGS.19678392.v1
25. Stosiak, M. (2011). Vibration insulation of hydraulic system control components. *Archives of Civil and Mechanical Engineering, 11*(1), 237–248.
26. Awad, H., & Parrondo, J. (2020). Hydrodynamic self-excited vibrations in leaking spherical valves with annular seal. *Alexandria Engineering Journal, 59*(3), 1515–1524. ISSN 1110-0168. https://doi.org/10.1016/j.aej.2020.03.033
27. Stosiak, M., Skačkauskas, P., Towarnicki, K., Deptuła, A., Deptuła, A. M., Prażnowski, K., Grzywacz, Ż., Karpenko, M., Urbanowicz, K., & Łapka, M. (2023c). Analysis of the impact of vibrations on a micro-hydraulic valve using a modified induction algorithm. *Machines, 11*, 184. https://doi.org/10.3390/machines11020184
28. Carpinlioglu, M. O., & Gundogdu M. Y. (2001). A critical review on pulsatile pipe flow studies directing towards future research topics. *Flow Measurement and Instrumentation, 12*, 163–174.
29. Adamkowski, A., & Lewandowski, M. (2006). Experimental examination of unsteady friction models for transient pipe flow simulation. *Journal of Fluids Engineering, ASME, 128*(6), 1351–1363.
30. Ferràs, D., Manso, P., Schleiss, A., & Covas, D. (2016). Experimental distinction of damping mechanisms during hydraulic transients in pipe flow. *Journal of Fluids and Structures, 66*, 424–446. ISSN 0889-9746. https://doi.org/10.1016/j.jfluidstructs.2016.06.009
31. Zarzycki, Z., Kudźma, S., Kudźma, Z., & Stosiak, M. (2007). Simulation of transient flows in a hydraulic system with a long liquid line. *Journal of Theoretical and Applied Mechanics, 45*(4), 853–871.

Application of Numerical Methods for Evaluating the Effect of Pulsating Flow on Hydraulic Line Vibration

3

The present chapter is dedicated to the utilization of numerical techniques to assess the impact of fluid flow pulsations on hydraulic line vibrations. Within this chapter, we delve into the mathematical representation of heightened pulsations within hydraulic system lines using the axial-piston pump model. Moreover, we expound upon the mathematical depiction of high-pressure hose behavior, which is contingent upon fluid flow. This chapter also introduces a fluid-solid coupling mathematical framework, encompassing mechanical equations resolved through the finite elements method (FEM), as well as hydrodynamic equations addressed via the method of characteristics (MoC). These two sets of equations are solved in tandem, yielding a comprehensive model designed to enhance our comprehension of the intricacies of fluid pulsations within hydraulic drive lines.

3.1 Modelling of Fluid Flow from an Axial-Piston Pump and Transfer of Pulsation to the Pipeline

The analysis encompasses the axial-piston pump in conjunction with the asynchronous motor (AM) and is denoted with pipeline connections. The fluid flow is characterized as one-dimensional and unsteady, with all local velocities taken into account. Furthermore, the fluid's motion is simplified to a one-dimensional representation, where local velocities are assumed to be equivalent to the average velocity despite the unsteadiness. The variables for velocity, $v(x, t)$, and pressure, $p(x, t)$, are contingent upon the longitudinal coordinate and time.

M. Stosiak and M. Karpenko, *Dynamics of Machines and Hydraulic Systems*, Synthesis Lectures on Mechanical Engineering, https://doi.org/10.1007/978-3-031-55525-1_3

3.1.1 Axial-Piston Mathematical Modelling

The kinetic energy of the pump's rotor is converted into the potential energy of the liquid within a high-pressure cylinder. To precisely simulate the interaction between the hydraulic pump piston and the liquid, specific scenarios of this interaction are examined. Each interaction scenario is influenced by the piston's behaviour within the cylinder as well as the computational grid employed. When the liquid interacts with the piston, the volumetric losses of the liquid are considered, as outlined by Aladjev and Bogdevicius [1].

Axial-piston pumps are extensively utilized in the hydraulic systems of mobile machinery due to their remarkable attributes, such as elevated output pressure, noteworthy efficiency, and dependable performance. In research applications, axial-piston pumps are paired with AMs to induce substantial fluid pulsations. The configuration employed for establishing the mathematical model of the motor-pump system is illustrated in Fig. 3.1. The dynamic model of the axial-piston pump, depicted in Fig. 3.2a, is constructed by delving into the operational principles and distinct structural characteristics of the pump. Within this model, each piston of the pump operates within a cylinder, engaging in close interaction with the liquid. The motion of each piston, contingent on the rotor's angular displacement φ_2, involves the expulsion (at high pressure) and suction (at low pressure) of the liquid. The kinematic representation, illustrated in Fig. 3.2b, outlines the configuration for designing the mathematical model of the axial-piston pump.

The dynamic model of the AM and axial-piston pump is analyzed as a complex hydraulic and mechanical system and is based on Karpenko and Bogdevičius [2]. In a comprehensive scenario, the AM model comprises a set of differential and algebraic equations. The torque exerted on the motor shaft is given by

$$\dot{M}_e = uc_v(w_o - u\dot{\varphi}_1) - d_v M_e, \tag{3.1}$$

where u—ratio between engine and motors shafts; c_v, d_v—engine parameters; w_o—angular velocity of the AM; $\dot{\varphi}_1$—angular velocity of AM shaft.

The inertia mass moment of motor shaft rotation is given by

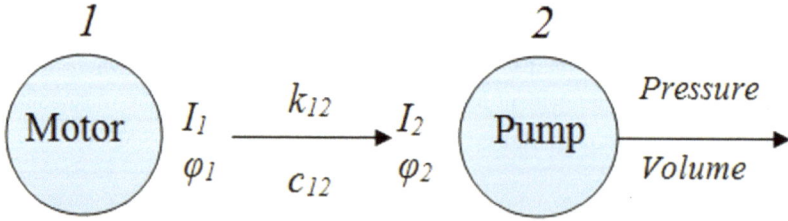

Fig. 3.1 Diagram for mathematical modeling of the motor-axial-piston pump

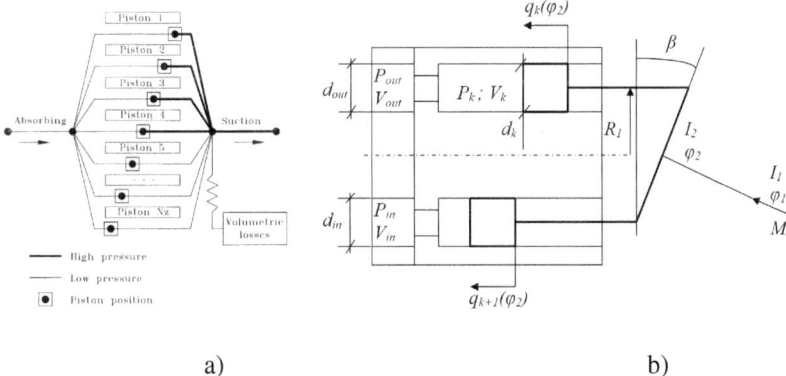

a)   b)

Fig. 3.2 Scheme for mathematical modeling: **a** hydraulic circuit of axial-piston pump; **b** Kinematic scheme of axial-piston pump

$$I_1\ddot{\varphi}_1 = M_e - k_{12}(\varphi_1 - \varphi_2) - c_{12}(\dot{\varphi}_1 - \dot{\varphi}_2), \tag{3.2}$$

where φ_1—angle of the AM shaft; $\ddot{\varphi}$—angular acceleration of AM shaft; $\dot{\varphi}_2$—angular velocity of axial piston pump shaft-plate; k_{12}, c_{12}—coefficients of stiffness and dampening; I_1—inertia moment of motor shaft.

The inertia mass moment of pump shaft-plate rotation is given by

$$I_2(\varphi_2)\ddot{\varphi}_2 = -\frac{1}{2}\frac{dI_2(\varphi_2)}{d\varphi_2}\dot{\varphi}_2^2 - k_{12}(\varphi_2 - \varphi_1) - c_{12}(\dot{\varphi}_2 - \dot{\varphi}_1) - M_{res}(\varphi_2, p_k), \tag{3.3}$$

where $I_2(\varphi_2)$—inertia moment of pump inlet shaft-plate; $\ddot{\varphi}_2$—angular acceleration of axial piston pump shaft-plate. M_{res}—torque moment resistance from the pump cylinders' operation:

$$M_{res}(\varphi_2, p_k) = \sum_{k=1}^{n_z} \frac{\pi d^2}{4} p_k R_1 \tan\beta \sin\alpha_k, \tag{3.4}$$

where n_z—number of pump pistons; R_1—radius of piston arrangement in the axial-piston pump; β—angle of inclination of the distributing washer; p_k—pressure in cylinder k of the pump; d—diameter of pump cylinders. α_k—angle of the pump piston's position:

$$\alpha_k = (k-1)\Delta\alpha + \varphi_2, \tag{3.5}$$

where $\Delta\alpha$—angle of pump piston initial position; k—number of the piston.

According to the construction of an axial-piston pump, every piston is coupled with the rotor, therefore, the interaction of piston and liquid flow in a cylinder, should be modeled in pair with equations of electrical engine movement (3.1, 3.2, 3.3, 3.4 and 3.5).

The displacement and its derivative of the pistons in the axial-piston pump block are defined as

$$q_k = R_1 \sin \beta (1 - \cos \alpha_k); \tag{3.6}$$

$$\dot{q}_k = \dot{\phi}_2 R_1 \sin \beta \sin \alpha_k. \tag{3.7}$$

Changes of volume (V_k) and its derivative in the piston chambers of the pump are defined as

$$V_k = A_c(L_c - q_k); \tag{3.8}$$

$$\dot{V}_k = -A_c \dot{q}_k = -\dot{\phi}_2 A_c R_1 \sin \beta \sin \alpha_k, \tag{3.9}$$

where A_c—the cross-sectional area of the pump piston chamber; L_c—length of the piston chamber of the pump.

The pressure changes in each cylinder of the pump when the pistons move are defined as

$$\dot{p}_k = \frac{K(p, \varepsilon)}{V_k(\phi_2)} \left[Q_{in} - Q_{out} - Q_{losses} - \frac{dV_k}{dt} \right], \tag{3.10}$$

where Q_{in}—intake volume of liquid in the piston chamber of the pump; Q_{out}—volume of fluid flowing out from the piston pump; Q_{losses}—volumetric losses.

The total change in inlet pressure (p_{in}) and output (p_{out}) of the axial-piston pump are defined as

$$\dot{p}_{in} = \frac{K(p, \varepsilon)}{V_{in}} \left[\sum_{k=1}^{n_z} Q_{ink} - \frac{dV_{in}}{dt} \right]; \tag{3.11}$$

$$\dot{p}_{out} = \frac{K(p, \varepsilon)}{V_{out}} \left[\sum_{k=1}^{n_z} Q_{outk} - \frac{dV_{out}}{dt} \right], \tag{3.12}$$

where V_{in}, V_{out}—volume of inlet and outlet pump chambers. Q_{ink}, Q_{outk}—inlet and outlet volume of liquid in the piston chambers of the pump:

$$Q_{ink} = A_{in}(\alpha_k)\mu_k(\alpha_k)\sqrt{\frac{2}{\rho}|p_k - p_{in}|} sign(p_k - p_{in}); \tag{3.13}$$

$$Q_{outk} = A_{out}(\alpha_k)\mu_k(\alpha_k)\sqrt{\frac{2}{\rho}|p_k - p_{out}|} sign(p_k - p_{out}), \tag{3.14}$$

where $A_{in}(\alpha_k)$, $A_{out}(\alpha_k)$—cross-sectional area of the piston with inlet and outlet chambers of the distributing washer; $\mu_k(\alpha_k)$—coefficient of hydraulic resistance (independent of cross-section surface); *sign*—an odd mathematical function that extracts the sign of a real number.

3.1.2   Numerical Simulation of Fluid Flow Within a Pipeline

The fluid motion is simplified as one-dimensional, where all local velocities are assumed to be equal to the average velocity, despite the unsteadiness. The velocity, $v(x, t)$, and pressure, $p(x, t)$, are functions of the longitudinal coordinate and time. This type of fluid movement is marked by waves of heightened and diminished pressure, emanating from the point of pressure variation across each cross-section of the vibrating fluid and causing deformation of the walls of high-pressure hoses. The equation describing fluid continuity in a differential form, as presented by Bogdevičius [3] and Karpenko [4], is as follows:

$$\frac{\partial}{\partial t}[S(x, t)\rho] + \frac{\partial}{\partial x}[S(x, t)\rho v] = 0, \tag{3.15}$$

where ρ, v—density and velocity of the fluid; $S(x)$ cross-section area of pipeline.

Fluid flow impulse (momentum) is defined as

$$\frac{\partial}{\partial t}[S(x, t)\rho v] + \frac{\partial}{\partial x}\left[S(x, t)\left(p + \rho v^2\right)\right] + \Pi(x)\tau + S(x, t)\rho a_x = p\frac{\partial S(x, t)}{\partial x}, \tag{3.16}$$

where $\Pi(x)$—the perimeter of cross-section; τ—tangential fluid stress in the inner surface of the pipeline; a_x—fluid acceleration along x axis.

Tangential fluid stress in the inner surface of the pipeline is determined as

$$\tau = \rho\lambda(\mathrm{Re}, \lambda)\frac{v|v|}{8}, \tag{3.17}$$

where λ—anti-turbulent fluid additives (friction coefficient).

$$\lambda(\mathrm{Re}) = \begin{cases} \frac{75}{\mathrm{Re}}, & \text{when } \mathrm{Re} \leq 2320 \\ \frac{0.3164}{\mathrm{Re}^{0.25}}, & \text{when } \mathrm{Re} > 2320 \end{cases}. \tag{3.18}$$

The system of Eqs. (3.15) and (3.16) can be written as a system of second-order quasi-linear differential equations:

$$[A]\left\{\frac{\partial u}{\partial t}\right\} + [B(u)]\left\{\frac{\partial u}{\partial x}\right\} = \{f(u)\}, \tag{3.19}$$

where $[A]$, $[B(u)]$ are matrices, and $\{f(u)\}$ is a vector that depends on t, x and elements of vector $\{u\}^T = [p, v]$.

$$[A] = \begin{bmatrix} 1 & 0 \\ 0 & 1 \end{bmatrix}; \ [B(u)] = \begin{bmatrix} v & c^2\rho \\ \frac{1}{\rho} & v \end{bmatrix}; \ \{f(u)\} = \left\{ \begin{array}{c} -\frac{\rho c^2}{r_0 + q_r} \dot{q}_r \\ \frac{2\tau}{\rho(r_0 + q_r)} - a_x \end{array} \right\}, \qquad (3.20)$$

where c—sound speed in the liquid with a certain amount of gas.

Setting the determinant of the matrix $[B(u)] - [A]\frac{dx}{dt}$ to zero gives an equation that allows the dx/dt derivative and characteristic direction to be determined. This equation has two various real roots $dx/dt = \lambda_i$ $(i = 1.2)$:

$$C^+: \frac{dx}{dt} = v + c; \ C^-: \frac{dx}{dt} = v - c. \qquad (3.21)$$

Sound velocity c in the liquid with a certain amount of gas, which is stored in the elastic hydraulic high-pressure hose, is equal to

$$c = \sqrt{\frac{K(p, \varepsilon)/\rho}{1 + \frac{K(p,\varepsilon) \cdot d}{E \cdot e} + \frac{\varepsilon}{\gamma}\left[\frac{K(p,\varepsilon)}{\gamma \cdot p} - 1\right]}}, \qquad (3.22)$$

where $K(p, \varepsilon)$—bulk modulus of elasticity of liquid; ρ—density of liquid; E—modulus of elasticity of the high-pressure hose; d—pipeline internal diameter; e—wall thickness of pipeline; γ—index of adiabatic process; ε—ratio between the gas volume in the liquid and the total volume of liquid (mixture).

The main idea of the characteristics method (MoC) is that the unknown variable velocity and pressure of the liquid at an instant moment in time $t + \Delta t$ is determined according to these parameters (Fig. 3.3).

Pressure and velocity at point D at any moment in time are determined by a system of nonlinear algebraic equations:

$$C^+ : \Phi_1 = v_D - v_L + \frac{1}{2}(p_D - p_L)\left[\left(\frac{1}{\rho c}\right)_L + \left(\frac{1}{\rho c}\right)_D\right]$$

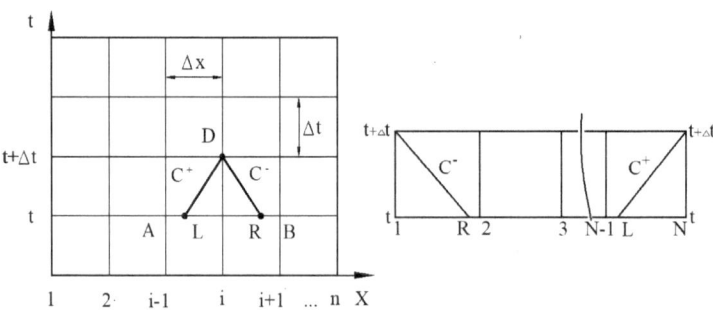

Fig. 3.3 Characteristic method scheme—including calculation scheme of the first and last points

$$-\frac{\Delta t}{2}\left[\left(\frac{f_1}{\rho c}\right)_L + \left(\frac{f_1}{\rho c}\right)_D\right] - \frac{\Delta t}{2}[(f_2)_L + (f_2)_D] = 0; \tag{3.23}$$

$$C^- : \Phi_2 = v_D - v_R + \frac{1}{2}(P_D - P_R)\left[\left(\frac{1}{\rho c}\right)_R + \left(\frac{1}{\rho c}\right)_D\right]$$

$$-\frac{\Delta t}{2}\left[\left(\frac{f_1}{\rho c}\right)_R + \left(\frac{f_1}{\rho c}\right)_D\right] - \frac{\Delta t}{2}[(f_2)_R + (f_2)_D] = 0 \tag{3.24}$$

where $f_1(p, v, q_r) = -\frac{\rho c^2}{r_0 + q_r(t)}\dot{q}_r(t)$; $f_2(v, p) = -\frac{2\tau}{\rho(r_0 + q_r)} - a_x$.

At points L and R, variables p and v are defined using Eqs. (3.23) and (3.24) according to the schemes in Fig. 3.3. This is expressed as a system of two nonlinear algebraic equations with unknowns p_L and v_L:

$$\Phi_3 = p_L - p_C - \theta(p_A - p_C)[v_L + c(p_L)] = 0; \tag{3.25}$$

$$\Phi_4 = v_L - v_C - \theta(v_A - v_C)[v_L + c(p_L)] = 0, \tag{3.26}$$

where $\theta = \Delta t/\Delta x$.

Analogously, the value of the variable R is set. In this case, we get a system of two nonlinear algebraic equations with unknowns p_R and v_R:

$$\Phi_5 = p_R - p_c + \theta(p_B - p_c)[v_R - c(p_R)] = 0; \tag{3.27}$$

$$\Phi_6 = v_R - v_c + \theta(v_B - v_c)[v_R - c(p_R)] = 0. \tag{3.28}$$

Differential equations of fluid movement in the pipeline are solved by the characteristics method (MoC). Pressure and velocity are determined by a system of nonlinear algebraic equations. The equations are solved by the Newton and Raphson method:

$$[J]_{i-1}\{\Delta Y\}_i = -\{\Phi\}_{i-1}; \quad \{Y\}_i = \{\Delta Y\}_{i-1} + \{\Delta Y\}_i; \tag{3.29}$$

$$[J]_{i-1} = \begin{bmatrix} \frac{\partial \Phi_{1,i-1}}{\partial p} & \frac{\partial \Phi_{1,i-1}}{\partial v} \\ \frac{\partial \Phi_{2,i\,1}}{\partial p} & \frac{\partial \Phi_{2,i\,1}}{\partial v} \end{bmatrix}; \{\Delta Y\}_i^T = [\Delta p_i, \Delta v_i]; \{\Phi\}_{i-1}^T = [\Phi_{1,i-1}, \Phi_{2,i-1}], \tag{3.30}$$

where $[J]_{i-1}$—Jacobian matrix; $\{\Delta Y\}_i$—vector d of unknown variables; i—iteration number. The proposed numerical model helps to describe and investigate fluid flow and pulsation inside hydraulic pipelines in combination with a hydraulic pump and other equipment (valves, throttles, etc.).

3.2   Mathematical Model of Pipeline Deformation Due to Fluid Flow

The mathematical model utilizes the FEM. Its fundamental concept involves partitioning the pipeline into discrete layers, resembling masses, interconnected by elements. These elements between the masses are characterized by a hybrid rheological model, combining aspects of the Maxwell and Kelvin–Voigt units, within a cohesive framework [5] and are presented in Fig. 3.4.

Each force between neighboring elements in a mathematical model can be described by formulas:

$$\ddot{F}_{0,1} + a_{10,1} \cdot \dot{F}_{0,1} + a_{10,1} \cdot F_{0,1} = b_{10,0}(q_1 - z)$$
$$+ b_{10,1}(\dot{q}_1 - \dot{z}) + b_{10,2}(\ddot{q}_1 - \ddot{z}), \ \ldots \ \ddot{F}_{n,n+1} + a_{n+1,1} \cdot \dot{F}_{n,n+1}$$
$$+ a_{n+1,0} \cdot F_{n,n+1} - b_{n+1,0}(q_{n+1} - q_n) -$$
$$- b_{n+1,1}(\dot{q}_{n+1} - \dot{q}_n) - b_{n+1,2}\left(\ddot{q}_{n+1} - \ddot{q}_n\right) = 0. \tag{3.31}$$

Coefficients of variables in the force equation are described by relationships between element coefficients of rheological models:

$$a_{i,0} = \frac{k_0 \cdot k_1 + k_0 \cdot k_2 + k_1 \cdot k_2}{c_1 \cdot c_2}; \ a_{i,1} = \frac{c_1 \cdot k_0 + c_1 \cdot k_2 + c_2 \cdot k_0 + c_2 \cdot k_1}{c_1 \cdot c_2};$$
$$b_{i,0} = \frac{k_0 \cdot k_1 \cdot k_2}{c_1 \cdot c_2}; \ b_{i,1} = \frac{c_1 \cdot k_0 \cdot k_2 + c_2 \cdot k_0 \cdot k_1}{c_1 \cdot c_2}; \ b_{i,2} = k_0. \tag{3.32}$$

A diagrammatic representation of the force and displacement between two adjacent masses (m_i and m_j) linked by the connecting element e_{ij} is illustrated in Fig. 3.5.

The changing position of i and j masses, in the global coordinate system of the model, can be defined by the following equations:

$$\{R_i\} = \{R_{i0}\} + \{q_i\}; \{R\}_j = \{R_{j0}\} + \{q_j\}, \tag{3.33}$$

where $\{R_{i0}\}$, $\{R_{j0}\}$—initial position of each mass in the global coordinate system; $\{q_i\}$, $\{q_j\}$—displacement of each mass by an external load.

The direction unit vectors $\{e_{ij}\}^T$, for defining the direction and value of vectors $\{q_i\}$ and $\{q_j\}$, are introduced:

Fig. 3.4 Rheological model of pipeline material element

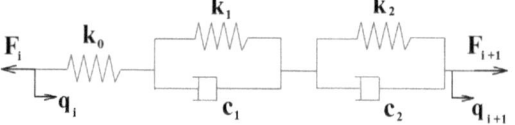

Fig. 3.5 Diagram for the investigation variables between two neighboring masses in the pipeline model

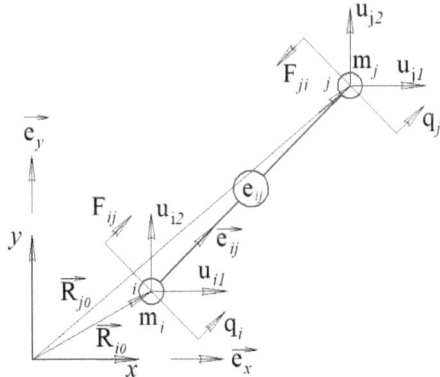

$$\{q_i\} = \{e_{ij}\}^T \{u_i\}; \{q_j\} = \{e_{ij}\}^T \{u_j\}, \tag{3.34}$$

where $\{u_i\}$, $\{u_j\}$ are the displacement vectors of m_i and m_j masses in the global coordinate system. They consist of horizontal ($\{u_{i1}\}$, $\{u_{j1}\}$) and vertical ($\{u_{i2}\}$, $\{u_{j2}\}$) displacements, respectively:

$$\{u_i\} = \begin{Bmatrix} u_{i1} \\ u_{i2} \end{Bmatrix}; \{u_j\} = \begin{Bmatrix} u_{j1} \\ u_{j2} \end{Bmatrix}. \tag{3.35}$$

The changing distance between m_i and m_j masses is defined by

$$\{R_{ij}\} = \{R_j\} - \{R_i\} = \{R_{j0}\} - \{R_{i0}\} + \{q_j\} - \{q_i\} = \{\Delta R_{ij0}\} + \{\Delta q_{ij}\}, \tag{3.36}$$

where $\{R_i\}$ and $\{R_j\}$—position of each mass in the global coordinate system; $\{\Delta R_{ij0}\}$—initial distance between masses of the model.

The vector of displacement $\{\Delta q_{ij}\}$ between m_i and m_j masses of the model is defined by

$$\Delta q_{ij} = \{q_j\} - \{q_i\} = \{e_{ij}\}^T \begin{Bmatrix} u_{j1} \\ u_{j2} \end{Bmatrix} - \{e_{ij}\}^T \begin{Bmatrix} u_{i1} \\ u_{i2} \end{Bmatrix} = \{e_{ij}\}^T \{\Delta q_{ij}\}. \tag{3.37}$$

The vector of velocity and acceleration between m_i and m_j masses of the model is defined by equations

$$\{\Delta \dot{q}_{ij}\} = \{e_{ij}\}^T (\{\dot{u}_j\} - \{\dot{u}_i\}) + \{\dot{e}_{ij}\}^T (\{u_j\} - \{u_i\});$$

$$\left\{\Delta \ddot{q}_{ij}\right\} = \{e_{ij}\}^T \left(\left\{\Delta \ddot{u}_{ij}\right\} + 2\{\dot{e}_{ij}\}^T \{\Delta \dot{u}_{ij}\}\right) + \left\{\ddot{e}_{ij}\right\}^T \{\Delta u_{ij}\}. \tag{3.38}$$

The direction unit vector $\{e_{ij}\}^T$ is defined by

$$\{e_{ij}\}^T = \frac{\{R_{ij}\}}{|\{R_{ij}\}|},$$ (3.39)

where $|\{R_{ij}\}| = \sqrt{\{R_{ij}\}^T \{R_{ij}\}}$—module of changing distance between m_i and m_j masses.

The velocity and acceleration of the direction unit vector are defined by the following equations:

$$\{\dot{e}_{ij}\}^T = \frac{|\{\dot{R}_{ij}\}||\{R_{ij}\}| - \{R_{ij}\}\frac{d}{dt}|\{R_{ij}\}|}{\{R_{ij}\}^T \{R_{ij}\}};$$

$$\{\ddot{e}_{ij}\}^T = \frac{\{\ddot{R}_{ij}\}}{|\{R_{ij}\}|} - 2\frac{\{\dot{R}_{ij}\}}{|\{R_{ij}\}|^3}\left(\{R_{ij}\}^T \{R_{ij}\}\right) -$$

$$- \frac{\{R_{ij}\}}{|\{R_{ij}\}|^3}\left(\{\ddot{R}_{ij}\}^T \{R_{ij}\} + \{\dot{R}_{ij}\}^T \{R_{ij}\} - 3\frac{\left(\{\dot{R}_{ij}\}^T \{R_{ij}\}\right)^2}{|\{R_{ij}\}|^2}\right),$$ (3.40)

where $\frac{d|\{R_{ij}\}|}{dt} = \frac{d}{dt}\sqrt{\{R_{ij}\}^T \{R_{ij}\}} = \frac{1}{|\{R_{ij}\}|}\{\dot{R}_{ij}\}^T \{R_{ij}\}$.

According to the mathematical model, the system of equations for m_i and m_j masses is defined as follows:

$$m_i[I]\begin{Bmatrix} \ddot{u}_{i1} \\ \ddot{u}_{i2} \end{Bmatrix} - \begin{bmatrix} \{e_x\}^T \\ \{e_y\}^T \end{bmatrix}\{F_{ij}\} = \begin{bmatrix} \{e_x\}^T \\ \{e_y\}^T \end{bmatrix}\{F_i\};$$

$$m_j[I]\begin{Bmatrix} \ddot{u}_{j1} \\ \ddot{u}_{j2} \end{Bmatrix} - \begin{bmatrix} \{e_x\}^T \\ \{e_y\}^T \end{bmatrix}\{F_{ij}\} = \begin{bmatrix} \{e_x\}^T \\ \{e_y\}^T \end{bmatrix}\{F_j\},$$ (3.41)

where $[I]$—the identity matrix, $[I] = \begin{bmatrix} 1 & 0 \\ 0 & 1 \end{bmatrix}$; $\{e_x\}^T$, $\{e_y\}^T$—vectors for determining the direction of forces on the x and y axes of the global coordinate system, respectively; F_{ij}—forces between m_i and m_j masses of the model; F_i and F_j—additional forces acting on m_i and m_j masses on the x and y axes of the global coordinate system, respectively:

$$\{F_i\}^T = \begin{bmatrix} F_{ix} & F_{iy} \end{bmatrix}; \{F_j\}^T = \begin{bmatrix} F_{jx} & F_{jy} \end{bmatrix}.$$ (3.42)

Each force between neighboring elements in the mathematical model can be described by the following equation:

$$\ddot{F}_{ij} + a_{1,ij}\dot{F}_{ij} + a_{0,ij}F_{ij} = b_{0,ij}\Delta q_{ij} + b_{1,ij}\Delta\dot{q}_{ij} + b_{2,ij}\Delta\ddot{q}_{ij}.$$ (3.43)

Or by applying Eqs. (3.37) and (3.38):

$$\ddot{F}_{ij} + a_{1,ij}\dot{F}_{ij} + a_{0,ij}F_{ij} = \left(b_{0,ij}\{e_{ij}\}^T + b_{1,ij}\{\dot{e}_{ij}\}^T + b_{2,ij}\{\ddot{e}_{ij}\}^T\right)\{\Delta q_{ij}\}$$
$$+ \left(b_{1,ij}\{e_{ij}\}^T + 2b_{2,ij}\{\dot{e}_{ij}\}^T\right)\{\Delta \dot{q}_{ij}\} + b_{2,ij}\{e_{ij}\}^T\{\Delta \ddot{q}_{ij}\},$$
$$(3.44)$$

where $a_{0,ij}$, $a_{1,ij}$, $b_{0,ij}$, $b_{1,ij}$, $b_{2,ij}$—coefficients that describe the relationships between element coefficients of rheological models, presented in Eq. (3.32).

The mechanical equation for m_i and m_j masses system is as follows:

$$\left[M_{ij}(q,\dot{q},\ddot{q})\right]\{\Delta \ddot{q}_{ij}\} + \left[C_{ij}(q,\dot{q},\ddot{q})\right]\{\Delta \dot{q}_{ij}\} + \left[K_{ij}(q,\dot{q},\ddot{q})\right]\{\Delta q_{ij}\} = \{B_{ij}(t,p,v)\},$$
$$(3.45)$$

where $[M_{ij}]$, $[C_{ij}]$, and $[K_{ij}]$ are matrices for variables' vector—$\left\{\Delta\ddot{q}\atop ij\right\}$; $\{\Delta\dot{q}_{ij}\}$; $\{\Delta q_{ij}\}$, respectively. The matrices $[M_{ij}]$, $[C_{ij}]$, and $[K_{ij}]$ are non-symmetrical.

An extended form of Eq. (3.45) is

$$\begin{bmatrix} m_i[I] & 0 & 0 \\ 0 & m_j[I] & 0 \\ b_{2,ij}\{e_{ij}\}^T & -b_{2,ij}\{e_{ij}\}^T & 1 \end{bmatrix}\begin{Bmatrix} \{\ddot{u}_i\} \\ \{\ddot{u}_j\} \\ \ddot{F}_{ij} \end{Bmatrix} +$$

$$+ \begin{bmatrix} 0 & 0 & 0 \\ 0 & 0 & 0 \\ b_{1,ij}\{e_{ij}\}^T + 2b_{2,ij}\{\dot{e}_{ij}\}^T & -b_{1,ij}\{e_{ij}\}^T - 2b_{2,ij}\{\dot{e}_{ij}\}^T & a_{1,ij} \end{bmatrix}\begin{Bmatrix} \{\dot{u}_i\} \\ \{\dot{u}_j\} \\ \dot{F}_{ij} \end{Bmatrix} +$$

$$+ \begin{bmatrix} 0 & 0 & -\{e_{ij}\} \\ 0 & 0 & \{e_{ij}\} \\ b_{0,ij}\{e_{ij}\}^T + b_{1,ij}\{\dot{e}_{ij}\}^T + b_{2,ij}\{\ddot{e}_{ij}\}^T & -b_{0,ij}\{e_{ij}\}^T - b_{1,ij}\{\dot{e}_{ij}\}^T - b_{2,ij}\{\ddot{e}_{ij}\}^T & a_{0,ij} \end{bmatrix}$$

$$\begin{Bmatrix} \{u_i\} \\ \{u_j\} \\ F_{ij} \end{Bmatrix} = \begin{Bmatrix} F_i \\ F_j \\ 0 \end{Bmatrix}.$$
$$(3.46)$$

In the model, the inner surface of the pipeline adopts a load from fluid pressure (Fig. 3.6a). This load uploads to masses as a distributed load on points loads and is described by $\{F_p\}$ forces:

$$F_p = \int_0^1 [N]^T p(\xi)2\pi r(\xi)L_e d\xi, \qquad (3.47)$$

where L_e—the distance between two bottom neighboring masses of the model; r—inner radius of pipeline; ξ—dimensionless coordinate, $\xi = (x - x_i)/(x_j - x_i)$; $p(\xi)$—fluid pressure for distributed load to point load.

Figure 3.6b shows a scheme to calculate the variable upload to two bottom neighboring masses of the pipeline model.

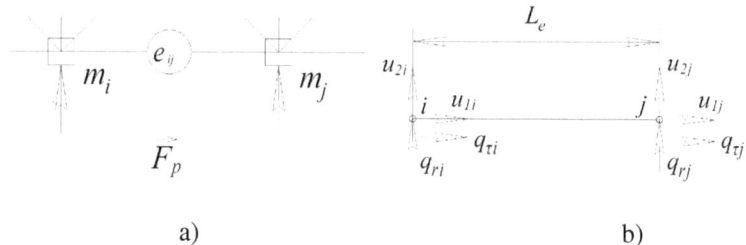

a)   b)

Fig. 3.6 Schemes for investigating the connection between hydraulic and mechanical components of pipeline modelling: **a** diagram depicting the distribution of fluid pressure load onto a point load applied to masses within the pipeline; **b** diagram outlining the computation of variable loading applied to the two lower adjacent masses

$$p(\xi) = p_i(1-\xi) + \xi p_j, \tag{3.48}$$

where p_i, p_j—fluid pressure upload to i and j masses, defined by modeling of fluid flow inside the pipeline.

The change of pipeline radius (of the inner layer), near the interaction between the pipeline inner surface and fluid, during modeling is defined by

$$r(\xi) = (1-\xi)(r_{i0} + u_{2i}) + \xi\left(r_{j0} + u_{2j}\right), \tag{3.49}$$

where r_{i0}, r_{j0}—initial radius value of i and j points, respectively; u_{2i}, u_{2j}—vertical displacement of i and j masses of the model by fluid pressure load.

The radial (q_r) and tangential (q_τ) displacements of pipeline elements depend on fluid flow and is defined by

$$q_r = u_{2i}(1-\xi) + u_{2j}\xi; \tag{3.50}$$

$$q_\tau = u_{1i}(1-\xi) + u_{1j}\xi. \tag{3.51}$$

The radial and tangential fluid stress in the inner surface of the pipeline, result from modelling of fluid flow inside pipeline. The friction forces between the pipeline inner surface and fluid flow can be described by load vector $\{F_\tau\}$:

$$\{F_\tau\} = \int_0^1 [N]^T \tau(\xi) 2\pi (r + q_\tau(\xi)) L_e d\xi, \tag{3.52}$$

where $\tau(\xi) = (1-\xi)\tau_i + \xi\tau_j$ and is defined by Eq. (3.17) by modeling the fluid flow inside the pipeline.

The total system of equations of the mechanical part of the pipeline has the following form:

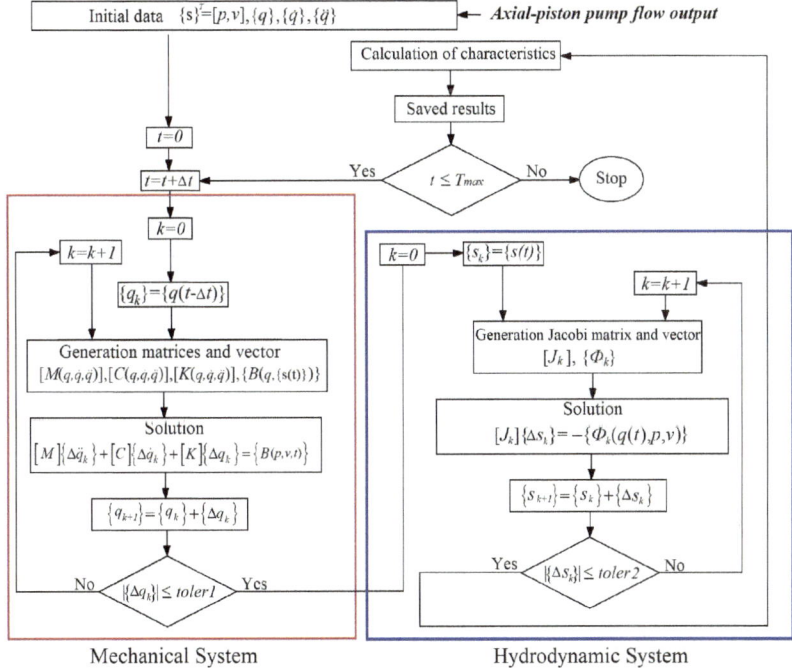

Fig. 3.7 The algorithm of calculation pipeline mechanical and hydrodynamic systems

$$[M]\left\{\ddot{Z}\right\} + [C]\{\dot{Z}\} + [K]\{Z\} = \{B(p, v, t)\}, \qquad (3.53)$$

where $\{Z\}^T = \left[\{q_1...q_n\}^T, \{F_1...F_{nf}\}^T\right]$.

The algorithm for calculating the pipeline mechanical system with a hydrodynamic system in couple is shown in Fig. 3.7.

3.3 Flow Pulsation and Pipeline Vibration Modelling Results

The interpretation and validation of axial-piston pump and pipeline modelling results are presented in the current sub-chapter.

3.3.1 Results of Axial-Piston Pump Modelling and Validation

Figure 3.8 shows the experimental setup designed for preliminary research focused on the analysis of fluid pulsation produced by an axial-piston pump. The key operational parameters of the research bench are as follows: axial-piston pump FOX 34—pressure

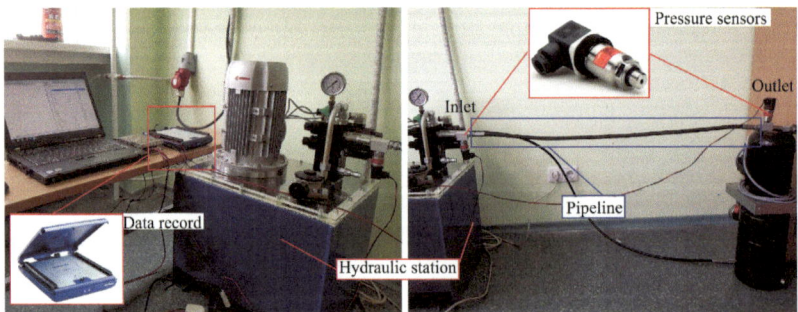

Fig. 3.8 Test bench used for mathematical model validation

in the system 2 MPa; flow rate approximately 24 dm^3/min; one-layer rubber pipeline diameter—1/2" (12.7 mm).

Based on the mathematical model, Fig. 3.9a, b show diagrams of the volumetric fluid flow rate and pressure pulsations at the outlet of the axial-piston pump. Additionally, spectra depicting the volumetric fluid flow rate and pressure pulsations at the axial-piston pump are shown in Fig. 3.10a, b. The constructed model of an axial-piston pump, operating at 3000 rpm, exhibits a nominal pressure at the output of 2.068 MPa and a flow rate of 5.6×10^{-4} m^3/s. Notably, the pump showcases significant fluid pulsations characterized by a frequency of 350 Hz, accompanied by a synchronous frequency of 50 Hz, as depicted in Fig. 3.10. The spectrum graphs also reveal harmonic frequencies with an incremental step of 350 Hz.

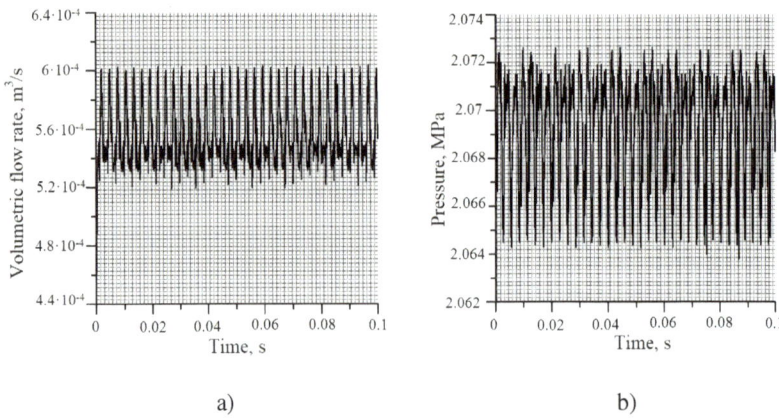

a)   b)

Fig. 3.9 Graphs illustrating the fluid flow at the pump outlet (from pump outlet to pipeline inlet): **a** graph depicting the fluid volumetric flow rate; **b** graph depicting fluid pressure pulsations

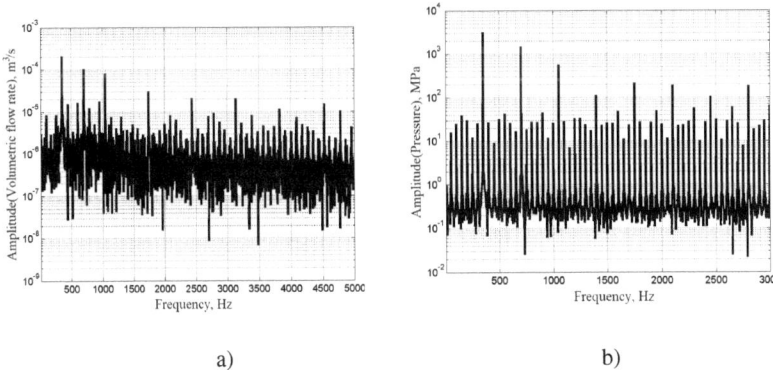

a)   b)

Fig. 3.10 Graphs of pressure pulsations at the pump outlet (from pump outlet to pipeline inlet): **a** logarithmic spectrum graph displaying the fluid flow rate; **b** logarithmic spectrum graph illustrating pressure pulsations

Utilizing measurements conducted on the research bench, Fig. 3.11a shows graphs depicting fluid pressure pulsations at the pump outlet (from pump to pipeline inlet). Additionally, a spectrum graph representing the fluid pressure pulsations of the axial-piston pump is depicted in Fig. 3.11b. Data recording took place over a duration of 10 s following the establishment of system processes. The fluid pressure pulsation graph showcases a periodicity of 0.1 s, selected to enhance pulsation visualization after reaching a stable load. Furthermore, the spectrum graph is presented up to a frequency of 500 Hz.

The developed model of an axial-piston pump, operating at 3000 rpm with an output pressure of 2 MPa, exhibits prominent fluid pulsations characterized by a primary frequency of 350 Hz and a synchronous frequency of 50 Hz. The mathematical model aligns

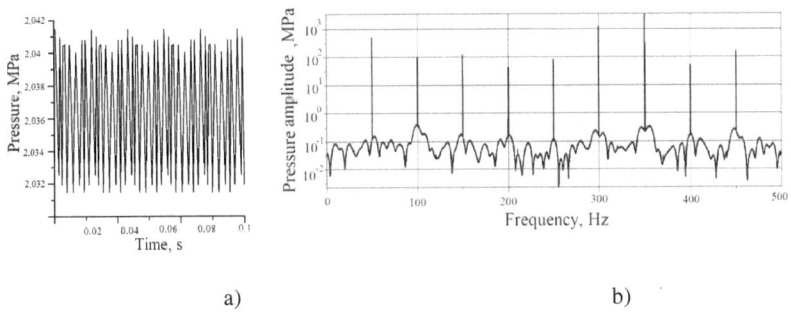

a)   b)

Fig. 3.11 Fluid pressure pulsations from measurements: **a** graphs illustrating pressure pulsations from the pump outlet to the pipeline inlet; **b** logarithmic graphs depicting pressure pulsations from the pump outlet to the pipeline inlet

with experimental data, reinforcing the observation that the pump operates at a fundamental synchronous frequency of 50 Hz when the motor rotates at approximately 3000 rpm. The hydraulic pressure within the system hovers around 2 MPa and is accompanied by substantial pulsations. This comprehensive mathematical model of the axial-piston pump encompasses factors such as wave motion of the liquid, localized pressure losses, volumetric fluid leakages, cavitation phenomena, and gaps within the mechanical drive. The devised technique has been effectively employed in research endeavours spanning volumetric hydraulic transmissions, pipeline deformation analysis, and frequency response modelling.

3.3.2 Results of Pipeline Modelling—Fluid Flow and Vibration

The obtained results for fluid pressure pulsation and fluid velocity inside the HPH are depicted in Fig. 3.12, which showcases a time period of 0.05 s following a stable load. This choice of time frame enhances the visualization of more pronounced pulsations.

Figure 3.13 illustrates alterations in radius on the inner surface of the pipeline caused by fluid pulsation. The outcome of spectrum analyses reveals the frequency response of velocity variations within the inner layer of the pipeline, as depicted in Fig. 3.14.

The analysis of the model reveals that the hydraulic pipeline undergoes deformation in response to fluid pressure pulsation. At the inlet (depicted in Fig. 3.12a), the fluid pressure stands at 2.07 MPa, accompanied by a fluid velocity of 4.4 m/s (Fig. 3.12b). Meanwhile, at the outlet, the pressure measures 2.06 MPa, and the fluid velocity is 4.33 m/s. Within the midpoint of the high-pressure hose (HPH), the fluid pressure reaches 2.065 MPa, accompanied by a fluid velocity of 4.36 m/s. The pressure and velocity signals at the inlet and outlet exhibit intricate waveforms, whereas at the midpoint, the forms are smoother. This

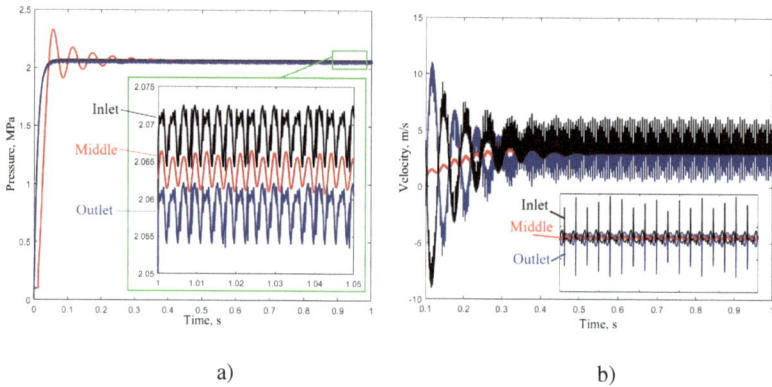

a)   b)

Fig. 3.12 Fluid flow: **a** pressure pulsation; **b** velocity of fluid

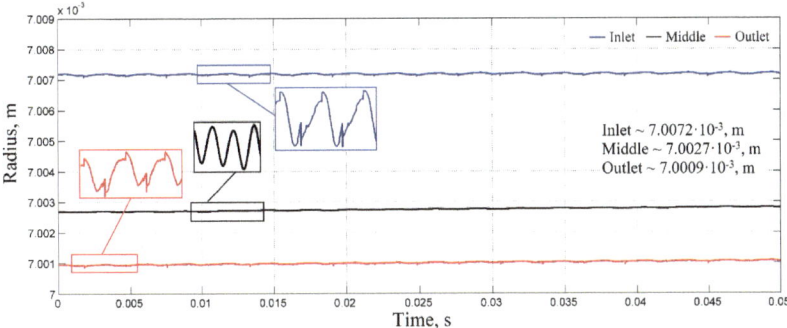

Fig. 3.13 Changes in radius along the inner surface of the pipeline due to fluid pulsation

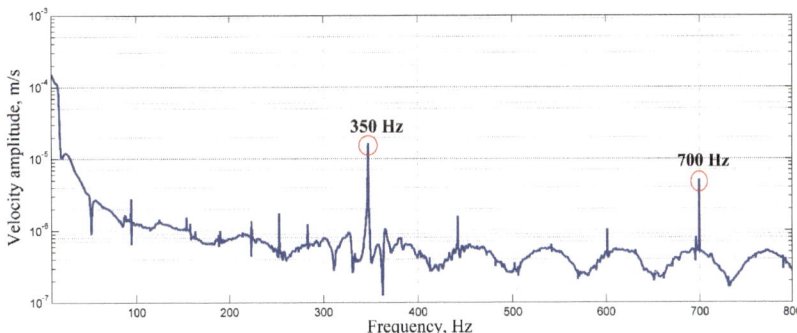

Fig. 3.14 Frequency response of velocity variations on the inner layer of the pipeline surface, plotted on a logarithmic scale

distinction arises from the application of the MoC to the hydraulic portion of the model. The deformation of the pipeline radius diminishes from the inlet (7.0072×10^{-3} m) to the outlet (7.0009×10^{-3} m), indicative of pressure drops along the entire length of the pipeline (Fig. 3.12a). This observation confirms that the viscous-elastic properties lead to attenuation of loads from the inlet to the outlet of the pipeline.

In the spectrum analysis of velocity changes along the inner layer of the pipeline surface, dominant amplitudes emerge at frequencies of 350 and 700 Hz (Fig. 3.14). These frequencies signify the transmission of effects from the fluid pulsation. This outcome further substantiates that loads from the fluid flow model are accurately conveyed from the hydrodynamic segment to the mechanical facet of the mathematical model.

References

1. Aladjev, V., Bogdevicius, M. (2006). *Maple: Programming, physical and engineering problems* (404 p.). Fultus Publishing.
2. Karpenko, M., Bogdevičius, M. (2018). Investigation of hydrodynamic processes in the system— "axial piston pumps—Pipeline—Fittings". In *Transport Problems 2018, X International Scientific Conference, VII International Symposium of Young Researchers: Proceedings* (pp. 832–843). Silesian University of Technology. ISBN 9788394571764.
3. Bogdevičius, M. (1991). Calculation of a non-stationary movement of a liquid in elastic-plastic and elastic-viscous-plastic pipelines. In *Processing of International Conference "Hydraulic machine"* (p. 48). Stuttgart University.
4. Karpenko, M. (2021). Investigation of energy efficiency of mobile machinery hydraulic drives. In *Dissertation* (164 p.). Vilnius Gediminas Technical University. https://doi.org/10.20334/2021-028-M
5. Bogdevičius, M., Karpenko, M., Bogdevičius, P. (2021). Determination of rheological model coefficients of pipeline composite material layers based on spectrum analysis and optimization. *Journal of Theoretical and Applied Mechanics, 59*(2), 265–278. https://doi.org/10.15632/jtam-pl/134802

This chapter presents known models of friction in hydraulic elements and the results of studies that identified transmission pathways of external mechanical vibrations to a control element of selected hydraulic valves. Firstly, a simplified physical system with a kinematically actuated pressure valve control element (ball or cone) will be analyzed. Then, we will focus on the impact of external mechanical vibrations on typical commercially available hydraulic valves. A modified mathematical model of the vibrating motion of the valve spool is proposed and experimentally verified while considering the theory of mixed friction.

4.1 Models of Friction in Hydraulic Drive and Control Elements

Several classic models of friction in pairs of motion that occur in typical hydraulic elements are found throughout the literature. The speed-dependent friction force model is most often used in hydraulic drives. The Hess-Soom model is among the basic models Chen [1]:

$$F_t = b \cdot v + F_{\text{mink}} \cdot sgn(v) + \frac{F_s - F_{\text{mink}}}{1 + \left(\frac{v}{v_s}\right)^2} \tag{4.1}$$

where F_s—rest friction force, F_{mink}—friction force corresponding to the minimum of the Stribeck curve, v—speed, v_s—experimental parameter in the speed dimension, b—experimentally determined parameter.

Pavelescu proposed a model in which the third term of the sum in Eq. (4.1) is changed [2, 3]:

$$F_t = b \cdot v + F_{mink} \cdot sgn(v) + (F_{sp} - F_{mink}) \cdot e^{-(v/v_s)^{\beta}} \tag{4.2}$$

where v_s and b are experimentally determined parameters.

An equally common model is the Tustin model:

$$F_t = b \cdot v + F_{mink} \cdot sgn(v) + F_{sk} \cdot e^{-(v/v_k)} \tag{4.3}$$

where v_k—the value of velocity resulting from the Stribeck characteristic at the transition to kinetic friction, F_{sk}—the difference between the force of rest and kinetic friction forces.

In cases where the linear model reflects the experiment well, the simplest model of the friction force resulting from the Newton formula is used:

$$F_{Tp} = b \cdot v \tag{4.4}$$

where b is the coefficient of viscous drag given by:

$$b = \frac{\mu \cdot A}{h} \tag{4.5}$$

μ—dynamic viscosity, A—contact area, h—gap height, oil film thickness.

In certain situations, in the description of friction processes, the coefficient of fluid friction can be found [3, 4]:

$$\eta = \frac{\mu \cdot v \cdot A}{F_N \cdot h} \tag{4.6}$$

resulting from the comparison of the Coulomb and Newton friction force models. F_N—downward force.

Literature Hol et al. [5] reports show a typical design node of a spool valve, in which the model of the friction force in the spool-sleeve pair can be described by Newton's law ignoring the Coulomb friction due to good lubricating conditions. Then, for the spool-sleeve pair, the friction force is described by the following relation [6]:

$$F_t = \pi \cdot d_t \cdot \frac{l}{h} \cdot \mu \cdot v \tag{4.7}$$

where d_t, l—diameter and length of the plunger, respectively, h—gap height, μ—dynamic viscosity of the working fluid.

However, practically, linear models based on Newton's law often do not adequately reflect the complex processes of friction. In many cases, to accurately represent the experience, it is necessary to use a total friction model that takes into account the sum of three forces [7]:

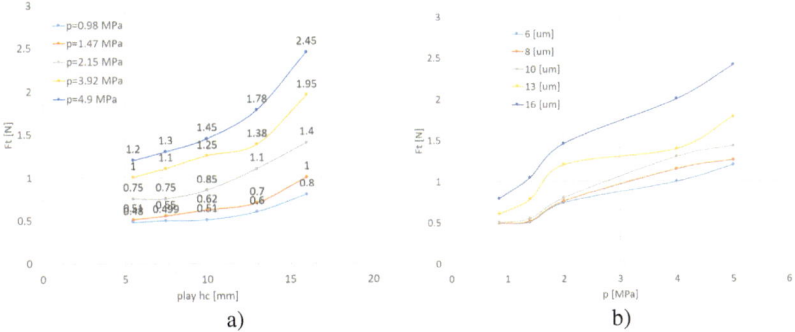

Fig. 4.1 The value of the friction force in the spool pair as a function of: **a** clearance for a fixed pressure; **b** pressure in the pair for a fixed clearance

$$F_t = F_{Tp} + F_{Ts} + F_p = b \cdot v + \left(F_{Ts} + F_{pmax} \cdot e^{-\sigma|v|} \right) \cdot sgn(v) \qquad (4.8)$$

where F_{Tp}—fluid friction force, F_{Ts}—dry friction force, F_p—adhesive force, F_{pmax}—maximum value of the adhesive force, σ—adhesion decay constant determined experimentally.

The following model of total friction forces has also been reported [8]:

$$F_t = b \cdot |v| + F_{Ts} sgn(v) + F_{od} \cdot \left(1 - \frac{|v|}{v_p} \right)^4 \qquad (4.9)$$

where F_{od}—detachment force, v_p—initial velocity of semi-fluid friction.

The value of the friction force F_t in the spool pair changes non-linearly as a function of the change in the pair's clearance h_c and the pressure p in the pair (Fig. 4.1).

For higher pressure values, the change in clearance h_c increases the value of the total friction force in the spool pair. The processes in the contact area are usually of a mixed nature: hydrodynamic and boundary. For this reason, the adequacy of a given model should be evaluated by comparing the experimental results.

4.2   Identification of the Contribution of Damping to the Transmission of Mechanical Vibrations to the Valve Control Element. Physical Model

The transmission pathways of external mechanical vibrations to the control element can be determined experimentally. The tests involve an electrohydraulic vibration inductor, on which a tank is mounted non-flexibly, and the valve control element (ball or poppet) is located on the spring, Fig. 4.2a, b. In order to determine the share of hydraulic oil in the transmission of vibration, considerations may apply when hydraulic oil is in the tank and when no hydraulic oil is in the tank. The test configuration consists of the following:

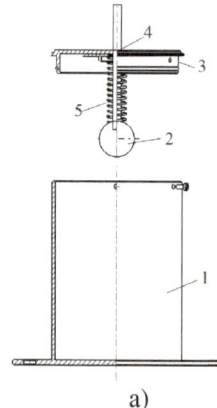

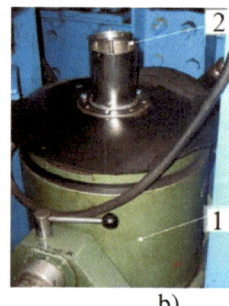

a)   b)

Fig. 4.2 Part view of measurement stand: **a** A part of the stand (physical model): 1—hydraulic oil tank, 2—valve control element (here in the form of a ball), 3—tank cover, 4—pin connecting the accelerometer with the valve control element being forced, 5—spring; **b** General view of the electric vibrator (1) with the hydraulic oil tank installed (2)

- the electrodynamic inductor of mechanical vibrations;
- a custom-made hydraulic oil tank with a spherical or cone-shaped element fixed on a spring, Figs. 4.2a and 4.3;
- measurement and data acquisition system (e.g., two accelerometers, a four-channel digital oscilloscope, a PC with specialized software).

The experiments employed to identify the pathways of the vibration transmission to the valve control element both measure and record the accelerations of vertical vibrations of the vibration exciter and the forced element. Table 3.1 lists the test results in which the forced element is a steel ball with mass m_k and radius r suspended on a spring with stiffness c_s and mass m_s, placed in hydraulic oil 68 at a temperature of 293 K. The parameters of the test system are listed in Table 4.1.

The spectra in Fig. 4.4a represent the deflection of the forced element from the equilibrium position under an externally acting force. To determine the contribution of oil to the transference of external mechanical vibrations to the forced element, the hydraulic

Fig. 4.3 A spatial model of a hydraulic oil tank test stand with a steel ball as theforced element

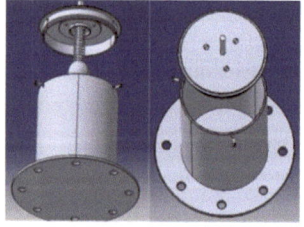

Table 4.1 The parameters of the testing system used for identifying the transmission pathways of external mechanical vibrations to an element immersed in oil and kinematically excited to vibrations

m_k [kg]	m_{st} [kg]	m_s [kg]	r_k [m]	r_{st} [m]	$\alpha_{st}/2$ [°]	μ [kg/m s]	g [m/s^2]	c_s [N/m]
0.0205	0.0195	0.005	0.0075	0.0075	45	0.18	9.81	875.62

where m_k—ball mass, m_{st}—cone mass, m_s—spring mass, r_k—ball radius, r_{st}—cone base radius, $\alpha_{st}/2$—half of the cone generatrix angle, μ—dynamic oil viscosity, g—gravitational acceleration, c_s—spring stiffness

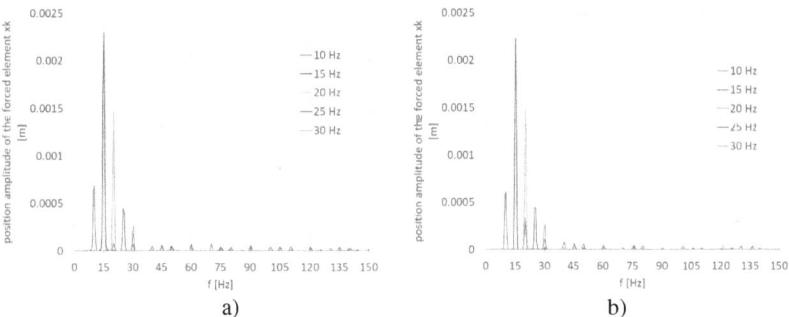

Fig. 4.4 The amplitude-frequency spectrum of vibrations of a kinematically forced ele-ment: **a** a steel ball immersed in system with hydraulic oil; **b** a steel ball in system without hydraulic oil

oil can be removed from the tank, and the experiment repeated for the same conditions of excitation. In this case, the forced element (steel ball with the characteristics listed in Table 4.1) is connected to the tank only by a spring with stiffness cs. The test results are presented in Fig. 4.4b in the form of an amplitude-frequency spectrum.

Analysis of the experimental results provided an amplitude-frequency spectrum of the kinematically forced steel ball and concluded that the maximum displacement amplitudes of the forced element occur for the excitation frequency $f = 15$ Hz. Similar conclusions can be drawn from theoretical considerations, with the mass of the forced element equal to the sum of the mass of the steel ball m_k, 1/3 of the mass of the spring m_s, and the mass of the accelerometer m_a measuring the vibration acceleration of the forced element. Therefore, the natural frequency of undamped vibrations can be expressed by the following equation:

$$\omega_0 = \sqrt{\frac{c_s}{m_k + \frac{1}{3}m_s + m_a}} = \sqrt{\frac{875,62}{0,0205 + 0,00167 + 0,11}} \approx 81,42 \ \frac{\text{rad}}{\text{s}} \quad (4.10)$$

The excitation frequency at which the excited element resonates is:

$$f_0 = \frac{\omega_0}{2 \cdot \pi} \approx 14\,\text{Hz} \quad (4.11)$$

Comparison of the plots from Fig. 4.4a, b can allow for the determination of the transmission pathway of external mechanical vibrations to the forced element (steel ball). Figure 4.5 shows a collective plot of the values of the ball vibration displacement amplitude as a function of the external excitation frequency, and the values of these amplitudes are summarized with percentage changes in Table 4.2.

The presented test results highlight that the main transmission pathway of external mechanical vibrations to the forced element in the considered physical model is a spring with stiffness c_s. In this case, the introduction of oil, and thereby taking into account damping k, did not fundamentally change the response of the forced system. This was due to the maximum relative difference Δx_w in the vibration amplitude of the forced element being approx. 11%.

Hydraulic valves are often used as a control element, called a poppet, in the form of a cone. In order to verify and refine the conclusions based on the tests of the steel ball (as a control element), similar tests have been carried out for the poppet as a control element in hydraulic lift valves. The tests consist of simultaneous measurements and recordings of vibration acceleration of the vibrator and the forced element. Figure 4.6 shows the

Fig. 4.5 Summary of the displacement amplitude of the oscillating motion of the ball in the test system, in which oil is present and absent in the system; f—frequency of ground vibrations

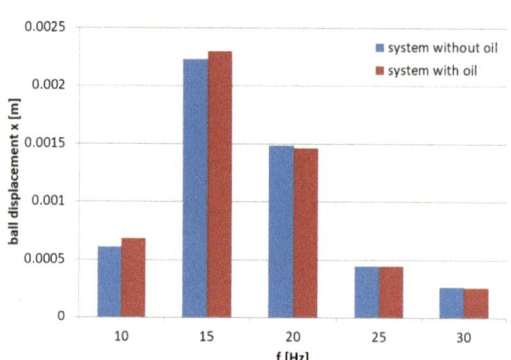

Table 4.2 Results of tests of the displacement amplitude of the oscillating motion of the ball in the test system, in which oil is present and absent in the system

f [Hz]	x_{bo} [m]	x_{zo} [m]	$\Delta x_w = \frac{|x_{zo} - x_{bo}|}{x_{zo}} \cdot 100\%$
10	0.000607	0.000682056	10.9589
15	0.002225	0.002294062	3.004044
20	0.001484	0.00146311	1.443614
25	0.000443	0.000443068	0.11486
30	0.000266	0.000258098	3.157895

x_{bo}—the amplitude of the steady-state vibrations of the system ball without hydraulic oil, x_{zo}—the amplitude of the steady-state vibrations of the system ball with hydraulic oil

Fig. 4.6 The
amplitude-frequency spectrum
of the vibration of a
kinematically forced
el-ement—a lift valve poppet
immersed in hydraulic oil

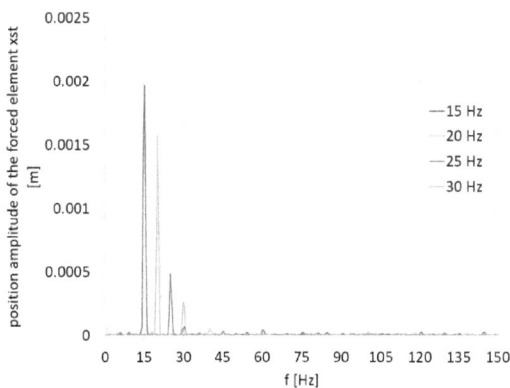

test results of when the forced element is a poppet with a mass m_{st}, base radius r_{st}, and
the angle of the cone generatrix α_{st} suspended on a spring with stiffness c_s and mass m_s
(Table 4.1). The oil tank to which the poppet is connected by the spring contains hydraulic
oil.

Furthermore, it can be observed that the maximum amplitude of displacement of the
forced poppet occurs at a frequency of external mechanical vibration of $f = 15$ Hz.
Therefore, it is reasonable to assume that, in this case, the natural vibration frequency
coincides with the excitation frequency. Based on the data listed in Table 4.1, the natural
frequency of undamped vibration can be determined for the considered element by the
following:

$$\omega_0 = \sqrt{\frac{c_s}{m_{st} + \frac{1}{3}m_s + m_a}} = \sqrt{\frac{875.62}{0.0195 + 0.00167 + 0.11}} \approx 81.67 \frac{\text{rad}}{\text{s}} \qquad (4.12)$$

The excitation frequency at which the element is excited can be:

$$f_0 = \frac{\omega_0}{2 \cdot \pi} \approx 14\,\text{Hz} \qquad (4.13)$$

When the oil is removed from the system and the tests repeated, the following results
are obtained (Fig. 4.7).

The collective graph showing the amplitude of the cone vibration displacement as a
function of the frequency of the external excitation is displayed in Fig. 4.8, where the
values of these amplitudes are compared with percentage changes in Table 4.3.

Comparison of the graphs from Figs. 4.6 and 4.7 allows for the determination of the
transmission pathway of external mechanical vibration to the forced element (poppet of
the hydraulic lift valve). The introduction of oil, where damping k is considered, did
not fundamentally change the response of the forced system in this testing, because the
maximum relative difference Δx_w in the vibration amplitude of the forced element is
approx. 17%.

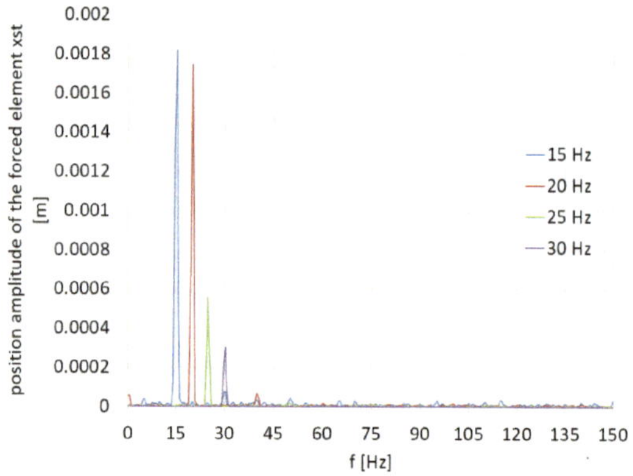

Fig. 4.7 The amplitude-frequency spectrum of the vibration of the kinematically forced element—the plug of the lift valve. System without hydraulic oil

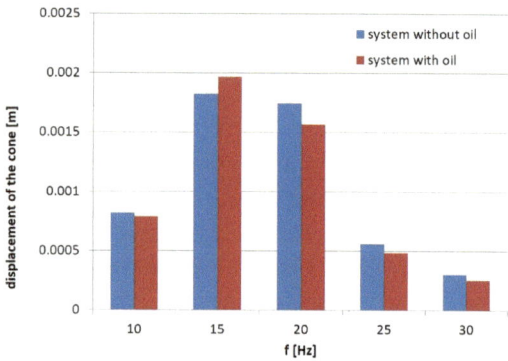

Fig. 4.8 The comparison of the displacement amplitude of the vibrating cone in the test system in which the system has and has not got oil; axis of abscissa—frequency of ground vibrations

Table 4.3 Cone vibration displacement as a function of the frequency of the external excitation—comparing with percentage changes

f [Hz]	x_{bo} [m]	x_{zo} [m]	$\Delta x_w = \dfrac{\lvert x_{zo} - x_{bo}\rvert}{x_{zo}} \cdot 100\%$
10	0.000819	0.000787	4.066074
15	0.001822	0.001967	7.371632
20	0.001746	0.001566	11.49425
25	0.000558	0.000485	15.05155
30	0.000301	0.000257	17.12062

x_{bo}—amplitude of the steady-state vibration of the cone in the system without hydraulic oil, x_{zo}—amplitude of the steady-state vibration of the system cone with hydraulic oil

As shown from the comparison of the results obtained for the system with hydraulic oil, the shape of the forced element influences its vibration caused by external mechanical vibration. A change in the shape of the forced element (from a ball to a cone) causes a drop in the vibration amplitude of the forced element in the resonance area. Also,

Fig. 4.9 Comparison of the displacement amplitude of the ball and cone vibrations as a function of the ground vibration frequency; system filled with hydraulic oil

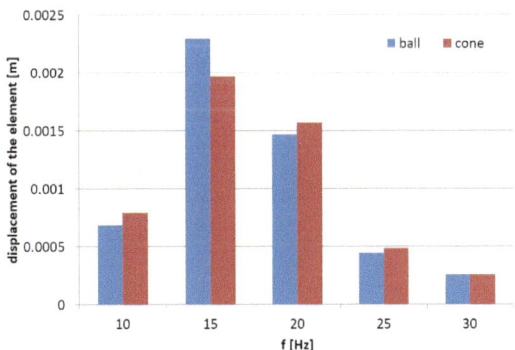

the difference in Δx_w increases, which indicates a greater contribution of damping to the transmission of vibration depending on the shape of the element excited to vibrate. Figure 4.9 compares the recorded vibration of a kinematically forced element (ball or poppet) immersed in hydraulic oil HL68 at a temperature of 293 K.

The strongest influence of the shape is observed for the frequency of $f = 15$ Hz (resonance without damping), and the amplitude decreases by approx. 15%.

4.3   Vibration of the Valve Control Element Caused by External Excitations

Experiments can demonstrate the influence of external mechanical vibrations on hydraulic valves. The linear hydrostatic drive simulator HydroPax ZY25 can be used as a source of external mechanical vibration. It is a hydraulic system, in which the essential element is the servo valve that controls the operation of the hydraulic actuator. The hydraulic actuator sets the position of the simulator table with a certain amplitude and frequency. The simulator reproduces real working conditions, Fig. 4.10. The hydraulic diagram of the simulator system is shown in Fig. 4.11a. The HydroPax ZY25 simulator consists of three main parts: hydraulic part; control device SYHCE1; control program HCE1.

To collect the relevant measurement data, a system was used to measure the following quantities:

- position of the hydraulic simulator table—potentiometric position sensor TLH-500,
- position of the control element of the tested element (in the case of a proportional spool valve)—spool position inductive sensor,
- pressure pulsation at the point of the presence of the tested element—piezoelectric pressure transducer from Piezotronics.

Fig. 4.10 Photograph of a
hydraulic simulator working as
a generator of mechanical
vibration that kinematically
excites the tested hydraulic
element

The measurement system enabled measurement, recording, processing in real-time, and saving the values on a hard disc of a connected PC. The arrangement of the measurement points is shown in Fig. 4.11b.

When a proportional directional valve is installed in a holder of the simulator table in such a way, the direction of the external vibration is parallel to the direction of movement of the spool in the sleeve (first series of tests) and in such a way that the direction of the external vibrations is perpendicular to the movement of the spool in the sleeve (series second test), by rotating the tested valve by 90°. A constant electrical control signal was applied to the coils of the proportional solenoids of the directional valve, causing the spool to deflect 2 mm from its neutral position. This makes it possible to overcome the positive rest overlap (the so-called dead zone of the directional valve). The flow rate of the working medium through the tested directional valve is 1×10^{-4} m^3/s (6 dm^3/min). The average pressure value at measurement point 5 (Fig. 4.11b) is 2 MPa. The over flow valve remained closed during the tests to eliminate additional components originating from working maximum valve 2 (as per Fig. 4.11b) from the spectrum. For clarity, a comparison study of the test results of three selected external excitation frequencies (table vibration): 40, 50, and 60 Hz is conducted in the form of amplitude-frequency spectra of pressure pulsation (measured at point 5, Fig. 4.11b) and displacement of the spool in the body of the tested proportional valve (p. 7, Fig. 4.11b)—Figs. 4.12, 4.13 and 4.14. In addition, Fig. 4.15 show the combined graphs of the amplitude-and-frequency spectrum of the spool vibration and pressure pulsations in the system (p. 5, Fig. 4.11b) for the direction of vibration parallel to the motion of the spool.

The presented vibration waveforms of the kinematically forced spool show that at the external excitation frequency of approx. 60 Hz, the amplitude of this vibration reaches the highest values. An even greater increase in the amplitude of the spool vibrations is observed compared to that of the excitation frequencies $f = 20$ and 60 Hz—the amplitude

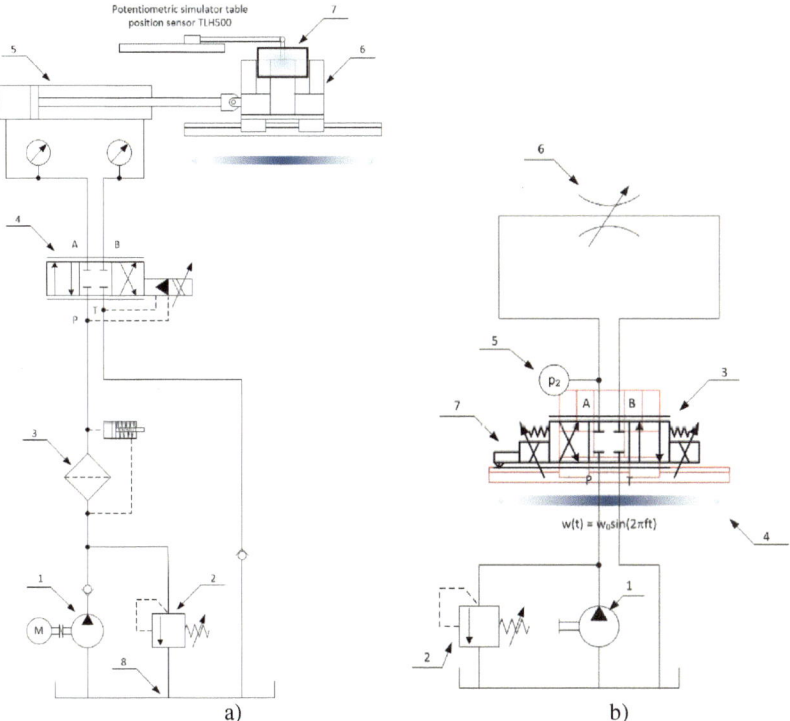

a)   b)

Fig. 4.11 Scemes in measurings: **a** hydraulic diagram of the linear hydrostatic drive simulator generating mechanical vibration: 1—feeding pump, 2—adjustable maximum valve, 3—oil filter, 4—servovalve 4WSE2EM10-45, 5—working actuator, 6—simulator table, 7—tested element: hydraulic lift valve, proportional spool valve, 8—hydraulic oil tank; **b** distribution of the measurement points in the test system of the proportional directional valve: 1—pump supplying the test system of the considered element, 2—overflow valve, 3—tested proportional directional valve 4WRE 6 E08-12/24Z4/M from Mannesmann-Rexroth placed on the simulator table, 4—hydraulic simulator table HYDROPAX ZY25, which kinematically excites the tested element according to a harmonic function and has the table position measuring point, 5—pressure change measurement point with the piezoelectric sensor M101A04 from Piezotronics, 6—adjustable throttle valve as a load of the test system, 7—measurement of the directional valve spool position with an inductive sensor

has an approx. 80% increase. At the same time, no vibration of the spool is observed in the absence of external mechanical vibration. According to the results, it can be concluded that the external mechanical vibrations acting on the valve body excite the vibrations of the valve control element, which is proportional to the spool in the presented case. The spool vibration induced in this way causes changes in the amplitude-frequency spectrum of the system's pressure pulsations. Harmonic components appear at frequencies corresponding to those of the spool vibration and the external excitation—Fig. 4.15b. The vibration transmission of the spool is also significantly influenced by the angle between

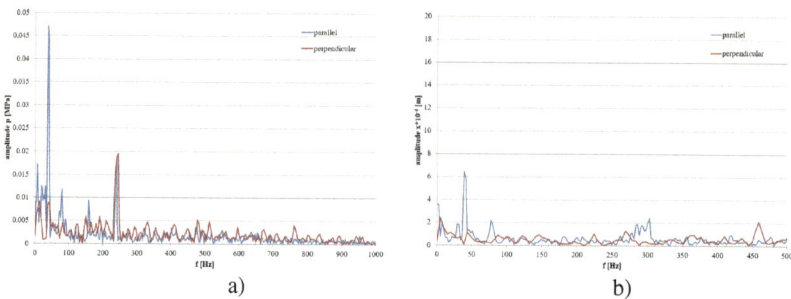

a)   b)

Fig. 4.12 The amplitude-and-frequency spectrum in a system with a vibrating proportional directional valve, where the vibration frequency of the simulator table is 40 Hz: **a** of pressure pulsation; **b** proportional valve spool vibrations

the direction of spool movement and the direction of external excitation (external vibration) [9, 10]. Two extreme cases should be considered, i.e., the direction of the spool movement that coincides with the direction of external vibration (angle $\alpha = 0°$) and the direction of spool movement perpendicular to the direction of external vibrations (angle $\alpha = 90°$). For the perpendicular direction (Figs. 4.12b, 4.13b, and 4.14b), the excitation of the spool vibrations is not observed. However, the spool vibrates in a parallel direction (angle $\alpha = 0°$). The dominant components of the spectrum of these vibrations correspond to the vibration frequency of the simulator table (frequencies of external excitations). The excited vibrations of the spool also cause periodic changes in the size of the proportional valve's throttling gaps, which result in periodical changes to the local resistance of the flowing working fluid. The spectrum of pressure pulsations contains components with frequencies corresponding to the excited vibration of the spool—Figs. 4.1a, b, 4.12a, and 4.13a. The pressure pulsation spectrum shows a component with a frequency of approx. 240 Hz resulting from the pulsation of the capacity of the positive displacement pump of the hydraulic system in which the tested directional valve operates (Fig. 4.11b).

4.4 Modeling and Experimental Verification of the Cooperation of the Spool Pair of the Hydraulic Directional Valve

This section presents the theoretical and experimental results related to the description of the nature of cooperation of the basic structural node of the spool valve, namely the spool-sleeve pair. This will allow further clarification of the transmission pathways description of the external mechanical vibration to the spool of the directional valve. Additionally, it will provide data to minimize the transmission of this vibration without enhancing the static and dynamic characteristics of the directional valve.

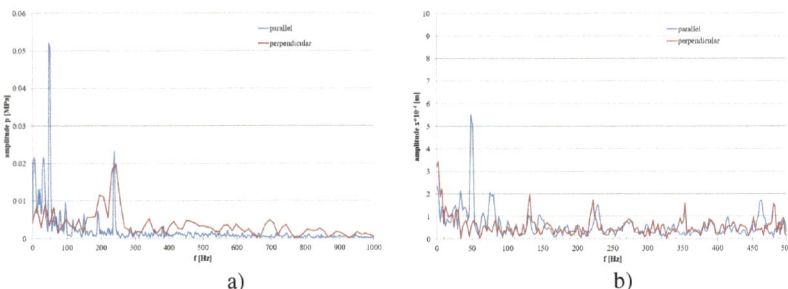

Fig. 4.13 The amplitude-and-frequency spectrum in a system with a vibrating proportional directional valve, where the vibration frequency of the simulator table is 50 Hz: **a** of pressure pulsation; **b** proportional valve spool vibrations

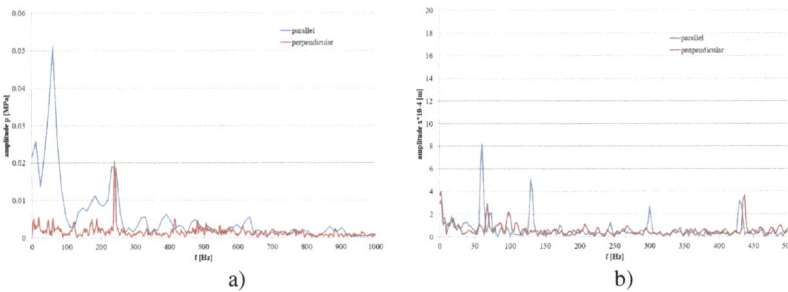

Fig. 4.14 The amplitude-and-frequency spectrum in a system with a vibrating proportional directional valve, where the vibration frequency of the simulator table is 60 Hz: **a** of pressure pulsation; **b** proportional valve spool vibrations

Before building the mathematical model of the spool valve movement, the following simplifying assumptions should be made:

- ignoring small influences;
- the tested system did not cause changes in the surrounding environment;
- using parameters concentrated in the points of the system instead of distributed as a function of length;
- linearization of relationships between the physical variables that describe the causes and effects;
- physical parameters are not functions of time;
- avoiding indeterminacy and ignoring the noise.

Based on the presented diagram, the following detailed simplifying assumptions should be considered in the mathematical model of the directional valve:

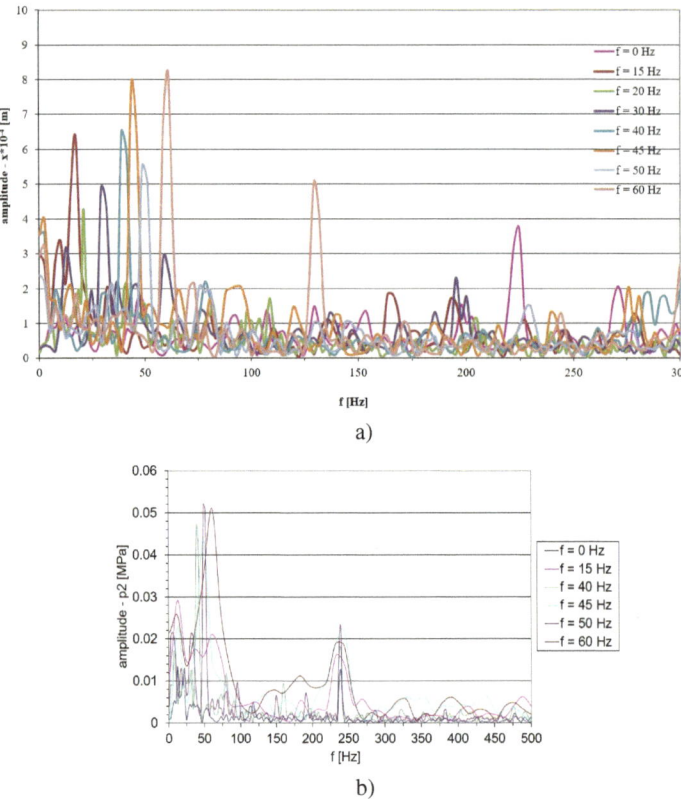

Fig. 4.15 The amplitude-and-frequency spectrum with a frequency of $f = 0$; 15; 20; 30; 40; 45; 50, and 60 Hz.: **a** proportional valve spool vibrations forced; **b** pressure pulsations

- assuming that the body of the directional valve is not deformed, neither is the spool, and the directional valve is attached to the vibrating base in a non-flexible manner;
- the holder in which the mounted directional valve is not subject to deformation;
- the mass of the fluid associated with the vibrating spool is omitted;
- assuming that in the hydraulic system of the tested directional valve, conditions are not conducive to the wave phenomena (there is no long hydraulic line), and thus models of systems with concentrated instead of distributed parameters are used for the description.

The working medium did not change its physical properties (the viscosity of the hydraulic oil does not change). Moreover, it is assumed that the spool is in the neutral position (no flow through the directional valve, therefore, no hydrodynamic force), it is not loaded

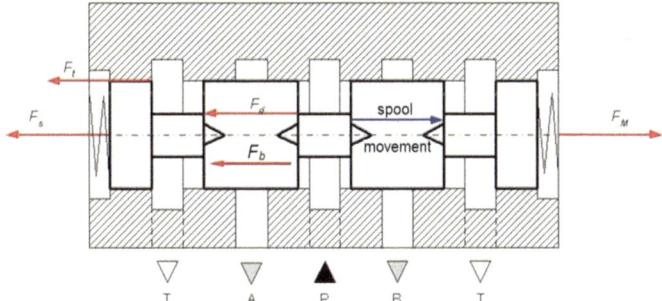

Fig. 4.16 Forces (force projection) acting on the spool of the directional valve equipped with pro-portional electromagnets: F_s—force from the stiffness of springs centering the spool, F_t—friction force in the spool pair, F_d—hydrodynamic force, F_b—inertia force of the spool, F_M—force from the controls (e.g., proportional solenoids)

with forces from static pressure, and there is no resistance forces movement from the friction of the seals.

When starting the analysis of the theoretical model of the spool valve, the starting point must be the balance of forces acting on the spool—Fig. 4.16.

The forces acting on the spool can be divided according to the direction of action, i.e., transverse and longitudinal. These forces determine the resistance of the spool movement that must be overcome when overriding. The lateral forces of the spool's movement relative to the sleeve do not have a direct effect but act perpendicularly to the axis of the spool. They determine the friction forces (and thus the resistance to movement) occurring on the spool's surface. However, the values of longitudinal loads are often much greater than those of the inertia and friction forces, therefore, they decisively determine the force necessary to override the spool of the directional valve.

The forces that act on the moving of the spool include:

- *spool inertia force*:

$$F_b = m_c \frac{d^2 x_{su}}{dt^2} \tag{4.14}$$

where: m_c—mass of movable adjusting elements of the piston spool and its associated liquid column, x_{su}—displacement of the spool.
- *force from the stiffness of the centering springs*, which is usually two centering springs located on both sides of the spool, with the same characteristics, forming a parallel connection:

$$F_s = c_{sz} \cdot x_{su} \tag{4.15}$$

c_{sz}—equivalent stiffness of the centering springs.

- the force of friction in the spool pair for friction given in a fluid friction model:

$$F_t = \pi \cdot d_t \cdot \frac{l}{h} \cdot \mu \cdot \frac{dx_{su}}{dt} \tag{4.16}$$

where, d_t, l—diameter and length of the piston, respectively, h—thickness of the gap in the spool pair.

Due to the good lubrication conditions of the spool pair, Coulomb friction can be neglected. However, such a simplification may lead to large discrepancies between the experimental results and those obtained based on a model for higher speeds of spool override [9, 10]. Based on the parameter for describing the friction processes [1, 2, 11], a model of mixed friction can be introduced to the mathematical model of forces acting on the vibrating spool of the directional valve.

The equation of forces (the projection of forces acting along the main axis of the spool) applied to the spool moving in the sleeve can be given for model 1:

$$m_c \frac{d^2 x_{su}}{dt^2} + c_{sz}(x_{su} - w) + \pi d_t \frac{l}{h} \mu \left(\frac{dx_{su}}{dt} - \frac{dw}{dt} \right) + k_t(d_{k0} - a_{k0}) \delta \left(\frac{dx_{su}}{dt} - \frac{dw}{dt} \right) = 0 \tag{4.17}$$

$$k_t = C_k \mu_s \tag{4.18}$$

where m_c—mass of the spool, associated fluid, mass of the sensor and 1/3 of the mass of the springs; x_{su}—displacement of the spool; c_{sz}—equivalent stiffness of the springs; d_t—diameter of the spool piston; l—length of the spool pistons; h—gap thickness in the spool-sleeve pair; μ—dynamic viscosity of the fluid; d_{k0}, a_{k0}—dimensions of the micro-wedge; k_t—coefficient of proportionality between the friction force and the value of the contact strain; C_k—contact stiffness of the surface; μ_s—coefficient of static friction.

The components of the sum of forces in Eq. (4.17) can be identified as follows: the first component of the sum is the inertia force of the spool movement (taking into account the mass of the spool and 1/3 of the mass of the centering springs), the second component of the sum describes the force from the equivalent stiffness of the springs that centers the spool in the body of the directional valve, the third term describes the fluid friction force in the spool pair that is proportional to the relative speed of movement of the spool in the valve body, and the fourth term describes the resting friction force that occurs when the spool changes the direction of its movement. Thus, the last two terms of the sum of forces in the model Eq. (4.17) describe the mixed friction force, which can be written as follows:

$$F_{Tm} = k_t(d_{k0} - a_{k0}) + \pi d_t \frac{l}{h} \mu \frac{dx_{su}}{dt} \tag{4.19}$$

The term $\delta\left(\frac{dx_{su}}{dt} - \frac{dw}{dt}\right)$ corresponds to the Dirac pseudo-function, taking the value of 1 when the relative velocity of the spool and sleeve is 0 and value 0 otherwise. This function can be approximated by a term of the following form:

$$\delta\left(\frac{dx_{su}}{dt} - \frac{dw}{dt}\right) \approx \frac{1}{1 + 10^{\theta} \cdot \left(\frac{dx_{su}}{dt} - \frac{dw}{dt}\right)^{\iota}} \tag{4.20}$$

The exponent θ should be as large as possible ($\theta \gg 1$) and the exponent ι should be both a large value and an even number ($\iota \gg 1$ and $\iota = 2 \times k_p$, where k_p belongs to the set of positive integers). These requirements are because, according to the considerations reported in Refs. Hutchings and Shipway [2], Chen [1], the denominator of Eq. (4.20) should tend to be infinity when the relative velocity of the spool and the sleeve is different from zero.

In addition, when the seals present in the spool pair are specific to such a design solution, the friction force should be included in the model (Eq. 4.17), and expressed in the following form [5, 6, 12]:

$$F_{Tu} = \pi \cdot D_p \cdot l_u \cdot f_t \cdot N_k + \pi \cdot D_p \cdot l_{u1} \cdot f_{t1} \cdot p \tag{4.21}$$

where l_u and l_{u1}—width of the sealing surface of the rubber ring and the retaining ring, respectively, f_t and f_{t1}—coefficient of friction between the rubber and the retaining ring material, respectively, D_p—inner diameter of the ring.

If there is no retaining ring, the second term of the sum in Eq. (4.21) should be neglected. The coefficient of the hydrodynamic friction, denoted as f_t, is expressed by the following formula [12, 13]:

$$f_t = c_p \frac{\mu \cdot v}{N_k \cdot h_0} \tag{4.22}$$

c_p—correction factor, N_k—value of contact pressures, h_0—thickness of the lubricating layer.

The cooperation of the spool-sleeve pair can also be represented by considering the mixed friction force as the sum of the molecular forces acting in the micro-areas of contact and the resistance forces in the liquid [1, 2]:

$$F_{Tm} = \mu_m \cdot N \tag{4.23}$$

$$\mu_m = C_1 \sqrt{\frac{dx_{su}}{dt} - \frac{dw}{dt}} + \frac{C_2}{\frac{dx_{su}}{dt} - \frac{dw}{dt}} \tag{4.24}$$

As shown, when the relative velocity of the spool is zero, the second term of the sum in Eq. (4.24) is undetermined. To avoid this, Eq. (4.24) can be modified so that the denominator of the second term of this equation is never equal to zero:

$$\mu_m = C_1 \sqrt{\frac{dx_{su}}{dt} - \frac{dw}{dt}} + \frac{C_2}{\frac{dx_{su}}{dt} - \frac{dw}{dt} + \gamma_m} \tag{4.25}$$

where γ_m is a constant close to zero. It is assumed that $\gamma_m = 1 \cdot 10^{-2}$. The constant γ_m has a unit consistent with that of velocity, that is [m/s].

Finally, the mixed friction model can be written as follows:

$$F_{Tm} = \left(C_1 \sqrt{\left| \frac{dx_{su}}{dt} - \frac{dw}{dt} \right|} + \frac{C_2}{\left| \frac{dx_{su}}{dt} - \frac{dw}{dt} \right| + \gamma_m} \right) \cdot N \tag{4.26}$$

where C_1 and C_2—constant values of the coefficients $\left[\sqrt{\frac{s}{m}} \right], \left[\frac{s}{m} \right]$, N—normal load [N], γ_m—a small value of a higher order $\left[\frac{m}{s} \right]$.

By considering the remaining forces acting on the moving spool of the directional valve, the following Eq. holds for model 2:

$$m_c \frac{d^2 x_{su}}{dt^2} + c_{sz}(x_{su} - w) + \pi d_t \frac{l}{h} \mu \left(\frac{dx_{su}}{dt} - \frac{dw}{dt} \right) +$$

$$+ sgn \left(\frac{dx_{su}}{dt} - \frac{dw}{dt} \right) \cdot \left(C_1 \sqrt{\left| \frac{dx_{su}}{dt} - \frac{dw}{dt} \right|} + \frac{C_2}{\left| \frac{dx_{su}}{dt} - \frac{dw}{dt} \right| + \gamma_m} \right) \cdot m_{su} \cdot g = 0 \tag{4.27}$$

where m_{su}—mass of the spool.

The proposed model of the valve spool movement, which considers the mixed friction (Eq. 4.27), describes both the case of the spool being at rest in relation to the sleeve and when these elements are in relative motion. The introduction of constant γ_m into Eq. (4.24), one should be cautious, as the value of the friction force increases to infinity when the spool is at rest relative to the valve body, which is an uncommon situation in practice.

The mixed friction model can be designed based on Coulomb's law and the equation of hydrodynamics [6, 10]. In that case, the force of mixed friction can be expressed by the following formula:

$$F_{Tm} = F_{Ts} - F_{Tp} \tag{4.28}$$

$$F_{Tm} = \mu_m \cdot N \tag{4.29}$$

$$\mu_m = \mu_0 - K_V \cdot \frac{\mu}{h} \cdot \frac{v}{p} \tag{4.30}$$

Therefore:

$$F_{Tm} = \left(\mu_0 - K_V \cdot \frac{\mu}{h} \cdot \frac{v}{p} \right) \cdot N \qquad (4.31)$$

where μ_0—dry friction coefficient; K_V—dimensionless characteristic of the contact geometry of surfaces exposed to friction; μ—dynamic viscosity of the liquid; [N s/m^2]; h—oil layer thickness, [m]; v—velocity of relative motion of friction surfaces, [m/s]; p = N/A$_t$, A$_t$—friction area, [m^2].

This equation should be included in the balance of the remaining forces acting on the moving spool—model 3:

$$m_c \frac{d^2 x_{su}}{dt^2} + c_{sz}(x_{su} - w) + \pi d_t \frac{l}{h} \mu \left(\frac{d x_{su}}{dt} - \frac{dw}{dt} \right) +$$

$$+ sgn\left(\frac{d x_{su}}{dt} - \frac{dw}{dt} \right) \cdot \left(\mu_0 - K_V \cdot \frac{\mu}{h} \cdot \frac{\left| \frac{d x_{su}}{dt} - \frac{dw}{dt} \right|}{p} \right) \cdot m_{su} \cdot g = 0 \qquad (4.32)$$

The coefficients of equations of mathematical models can be parametrized through experiments and with the use of computational tools for multi-parameter estimation.

If the presence of mixed friction is neglected due to good lubrication conditions in the spool pair and only liquid friction is assumed, then the description of the movement of the spool can be given by the following equation for model 4:

$$m_c \frac{d^2 x_{su}}{dt^2} + c_{sz}(x_{su} - w) + \pi d_t \frac{l}{h} \mu \left(\frac{d x_{su}}{dt} - \frac{dw}{dt} \right) = 0 \qquad (4.33)$$

4.4.1   Experimental Testing

The conducted experimental tests consisted of the main stage of the 4/3 directional valve (type 4WEH 10 J46/6EG24NETK4/B10) made by Bosch-Rexroth, being subjected to external mechanical vibrations with a frequency of 10 to 100 Hz. During the tests, the control chambers of the main directional valve stage were unloaded, i.e., their pressure was equal to atmospheric pressure. The waveform of the vibration acceleration of the directional valve spool and of the body was recorded. The tests involved a directional valve flooded with HL68 or AzollaZS22 oil, at a temperature of 20 °C (with a dynamic viscosity of 612 × 10^{-4} or 198 × 10^{-4} N s/m, respectively). After a series of tests, the pair of centering springs were changed. Spring stiffness and oil type are given in Table 4.4. The mechanical vibration originated from the Hydropax ZY-25 linear hydrostatic drive simulator. The direction of forced vibration was the same as the direction of movement of the spool in the sleeve.

Table 4.4 Selected parameters of the model of the tested directional valve

No	Spring stiffness c_S [N/m]	Oil name
1	816	AzollaZS22
2	2923	HL68
3	2923	AzollaZS22
4	816	HL68

The tested directional valve was mounted in a non-flexible manner in a special holder of the simulator table. Figure 4.17a shows the directional valve during the tests. Figures 4.17b and 4.18 show the diagram of the directional valve and the measurement circuit.

The experimental test results are presented using a bar chart. The acceleration amplitude of the spool and body vibrations for the selected frequency of excitation (vibration

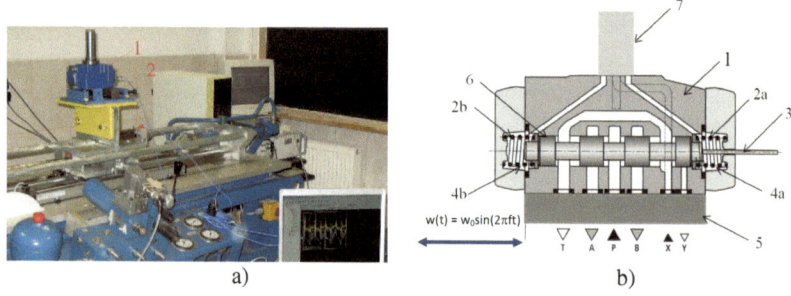

a)   b)

Fig. 4.17 The tested directional valve: **a** directional valve tested in the simulator table holder: 1—directional valve tested; 2—simulator table holder; **b** tested directional valve: 1—directional valve body; 2a, 2b—centering springs; 3—pin measuring the spool vibration acceleration; 4a, 4b—spool control chambers; 5—mounting plate; 6—spool; 7—oil filling tank

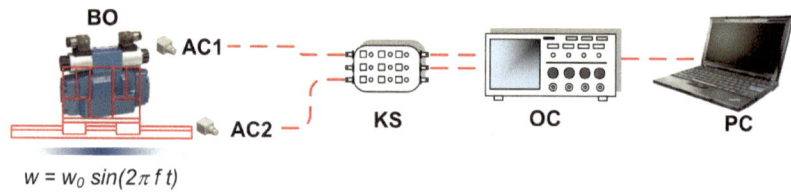

$w = w_0\ sin(2\pi f t)$

Fig. 4.18 Diagram of the measurement circuit for mechanical vibration acceleration: BO—item tested—main stage of the directional valve 4WEH J46/6EG24NETK4/B10, AC1—accelerometer measuring the vibration accelera-tion of the directional valve spool, AC2—accelerometer measuring the acceler-ation of vibration of the directional valve body, KS—signal conditioner Vib-Amp PA3000, OC—digital oscilloscope TDS224 Tektronix, PC—personal computer

of the simulator table) were compared (Figs. 4.19 and 4.20). The replacement was applied to the centering springs and the hydraulic oil with which the directional valve was filled and the spool moved.

The above experimental results indicate that the movement of the spool follows the movement of the vibrating body. In other words, an increase in the amplitudes of body vibrations coincides with an increase in the amplitude of the vibrations of the spool for the same frequencies. The use of springs of different stiffnesses in the tests leads to the vibrating spool acquiring different values of undamped vibration natural frequency. For the equivalent stiffness of the springs equal to 1632 N/m, the natural frequency of the spool's undamped vibration is approx. 15 Hz, and when the equivalent stiffness of the springs is 5846 N/m, this value is approx. 29 Hz. Figures 4.20a, b show a slight increase in the amplitude of vibration acceleration in the resonance area of the spool. However, for series 1 (Fig. 4.19a), an increase in the amplitude of the spool vibration acceleration in the resonance area is observed. In series 4 (Fig. 4.19b), this increase is strongly suppressed

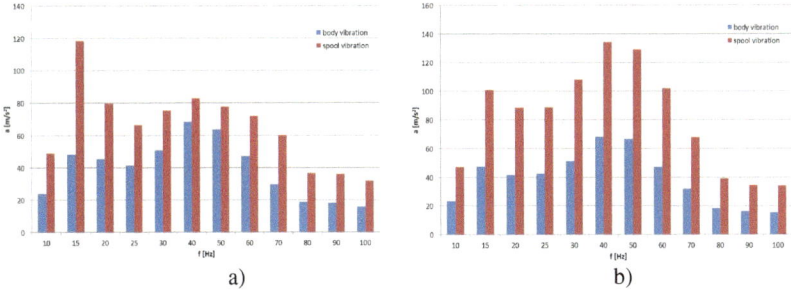

Fig. 4.19 The amplitude of the vibration acceleration of the directional valve body and spool as a function of the exciting vibration frequency; equivalent spring stiffness 1632 N/m: **a** AzollaZS22 oil; **b** HL68 oil

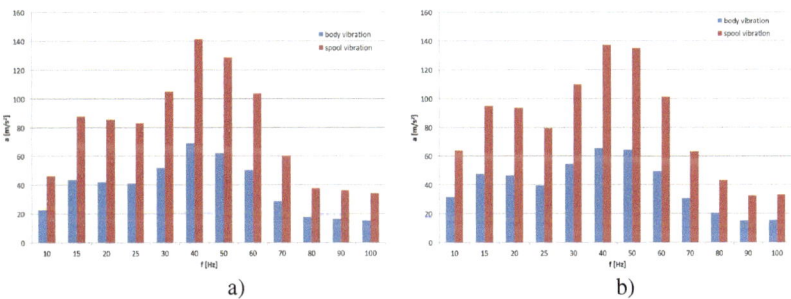

Fig. 4.20 The amplitude of the vibration acceleration of the directional valve body and spool as a function of the exciting vibration frequency; equivalent spring stiffness 5846 N/m: **a** AzollaZS22 oil; **b** HL68 oil

by the oil with a higher dynamic viscosity (HL68 instead of AzollaZS22—dynamic oil viscosity: 612×10^{-4} and 198×10^{-4} N s/m, respectively).

4.4.2 Estimation of Parameters of Mathematical Models

When considering the mixed friction, the coefficients of the model of the spool's vibrating motion can be parametrized based on the experimental results and with a specialized computer tool—Matlab/Simulink Parameter Estimation. Table 4.5 lists the most important parameters of the spool pair.

Using the Simulink Parameter Estimation tool, the values of the parameters of the mixed friction model can be obtained using Eq. (4.17), Table 4.6.

Using the Simulink Parameter Estimation tool, the values of the parameters of the mixed friction model can be obtained using Eq. (4.27), Table 4.7.

A similar procedure was used for the estimation of the parameter values of model 3, described by Eq. (4.32), Table 4.8.

Table 4.5 Selected parameters of the spool pair

Designation	Name	Value	Unit
m_c	Mass of spool + mass of associated liquid + 1/3 mass of springs	0.18	[kg]
c_{sz}	Equivalent spring stiffness	1632 or 5846	[N/m]
m_{su}	Spool mass	0.172	[kg]
d_t	Piston diameter	20×10^{-3}	[m]
h	Radial clearance between the spool and sleeve	15×10^{-6}	[m]
l	Piston length	32×10^{-3}	[m]
μ	Dynamic oil viscosity (HL68 or AzollaZS22 at temperature 20 °C)	612×10^{-4} or 198×10^{-4}	[N s/m]

Table 4.6 Values of estimated parameters of model 1

Equivalent spring stiffness [N/m], hydraulic oil	a_{k0} [m]	d_{k0} [m]	k_t [N/m]	Objective function
1632, HL68	0.00356	0.00565	46.86	4.1084×10^7
5846, HL68	1.9854×10^{-6}	0.00732	100	7.618×10^7
5846, AzollaZS22	2×10^{-6}	8×10^{-6}	100	1.0015×10^8
1632, AzollaZS22	7.1717×10^{-5}	0.0032	99.985	3.1135×10^7

Table 4.7   Values of estimated parameters of model 2

Equivalent spring stiffness [N/m], hydraulic oil	$C_1 \left[\sqrt{\frac{s}{m}}\right]$	$C_2 \left[\frac{s}{m}\right]$	Objective function
1632, HL68	10.0536	9.6365	1.67×10^7
5846, HL68	9.9998	9.9906	1.81×10^7
5846, AzollaZS22	9.9998	10.0046	1.72×10^7
1632, AzollaZS22	9.9684	15.9903	8.98×10^6

Table 4.8   Values of estimated parameters of model 3

Equivalent spring stiffness N/m, hydraulic oil	μ_0	K_V	Objective function
1632, HL68	0.16025	0.77652	2.4495×10^7
5846, HL68	0.10885	−0.047217	7.4157×10^7
5846, AzollaZS22	0.10039	−0.0008	9.8865×10^7
1632, AzollaZS22	0.03823	0.025445	3.1519×10^7

If mixed friction is omitted and the spool is assumed to work with the sleeve only in the conditions of the fluid friction expressed by Eq. (4.16), and if the values of the model parameters of Eq. (4.33) are estimated, then the parameter values listed in Table 4.9 are obtained.

The minimization of the value of the objective function is used to evaluate the mathematical models of the movement of the vibrating spool. The objective function is understood as a weighted sum of squared errors [14].

As an additional criterion, the correctness of the representation of the spool movement is the ratio of the vibration acceleration amplitudes after the estimation and from the experiment a_{est}/a_{pom}, which tends to be 1. The average values of the objective function and the ratio a_{est}/a_{pom} are presented in Table 4.10. Additionally, the value of the ratio of the amplitudes of vibration acceleration after estimation is considered, and from the experiment a_{est}/a_{pom}, which tends to be 1. The frequency-averaged values of the objective function and the ratio a_{est}/a_{pom} are presented in Table 4.10.

Table 4.9   Values of estimated parameters of model 4

Equivalent spring stiffness [N/m], hydraulic oil	h[m]	Objective function
1632, HL68	3.1171×10^{-5}	2.5752×10^7
5846, HL68	8.4188×10^{-6}	2.9586×10^7
5846, AzollaZS22	1.9375×10^{-5}	$2.2014e \times 10^7$
1632, AzollaZS22	8.7237×10^{-5}	$2.9153e \times 10^7$

Table 4.10 Average values of the objective function and acceleration amplitude ratios for individual models

	Model 1 (4.17)	Model 2 (4.27)	Model 3 (4.32)	Model 4 (4.33)
(Objective function)$_{avg}$	6.27×10^7	1.52×10^7	5.73×10^7	2.66×10^7
(a_{est}/a_{pom})$_{avg}$	0.65	0.89	0.70	0.40

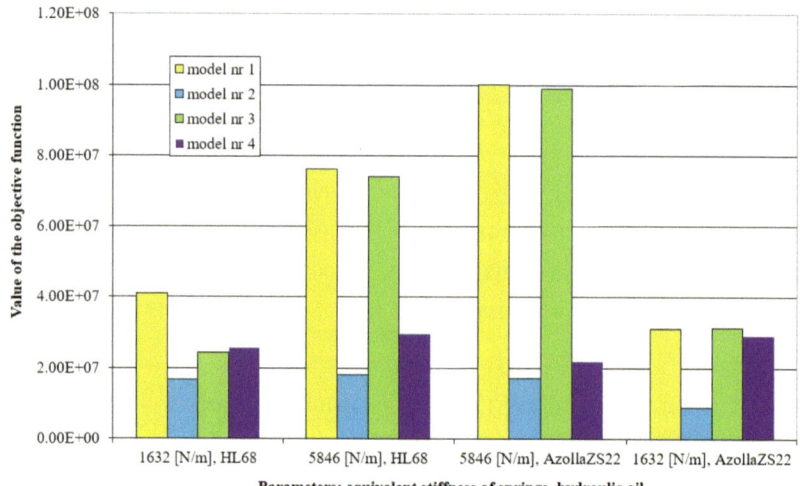

Fig. 4.21 The value of the objective function for the considered models and given pa-rameters: equivalent stiffness of springs and hydraulic oil

Figure 4.21 shows the comparison of the objective function values for equal values of the equivalent stiffness of springs and oils of different viscosities for models 1–4.

The conducted experiments and computer simulations, which estimated the parameters of the model equations, allow for the comparison of the time waveforms of the spool vibrations—Figs. 4.22, 4.23, 4.24 and 4.25. Four previously presented models are compared: three mixed friction models and a liquid friction model (model 4), where the equivalent spring stiffness is 1632 N/m, and the directional valve is filled with HL68 oil at 20 °C.

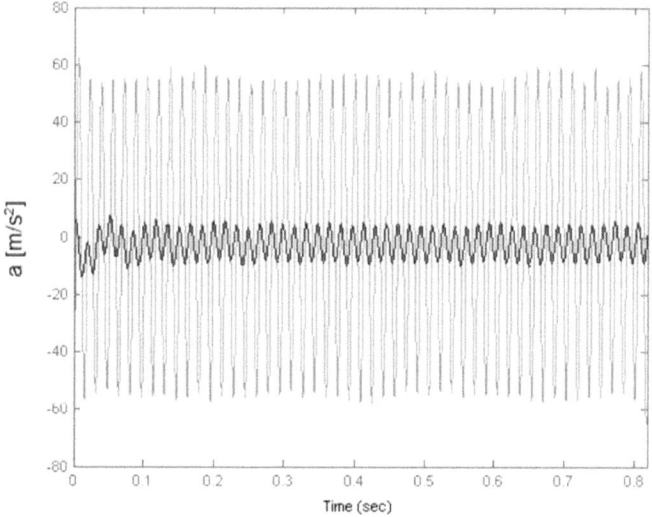

Fig. 4.22 The time waveform of the acceleration of the directional valve spool vibrations for the excitation frequency of 60 Hz. Blue color—solution after estimation of parameters of model 1 (4.17), gray color—results of experimental measurements

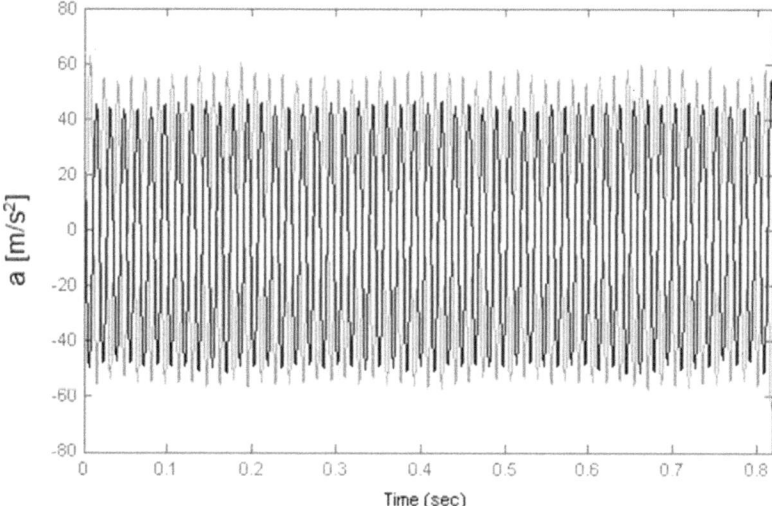

Fig. 4.23 The time waveform of the acceleration of the directional valve spool vibrations for the excitation frequency of 60 Hz. Blue color—solution after estimation of parameters of model 2 (4.27), gray color—results of experimental measurements

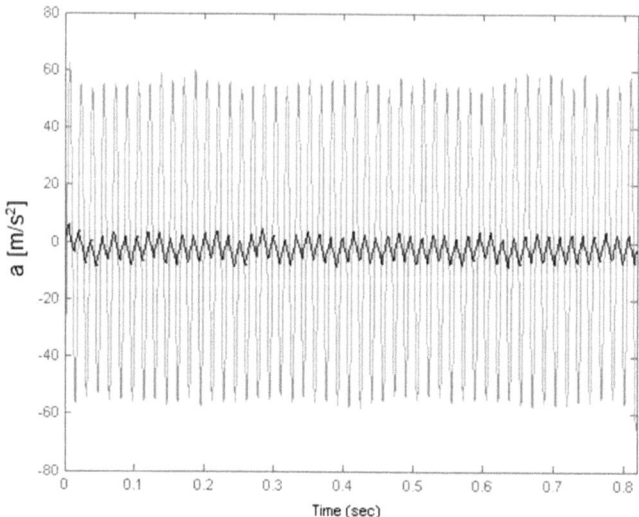

Fig. 4.24 The time waveform of the acceleration of the directional valve spool vibrations for the excitation frequency of 60 Hz. Blue color—solution after estimation of parameters of model 3 (4.32), gray color—results of experimental measurements

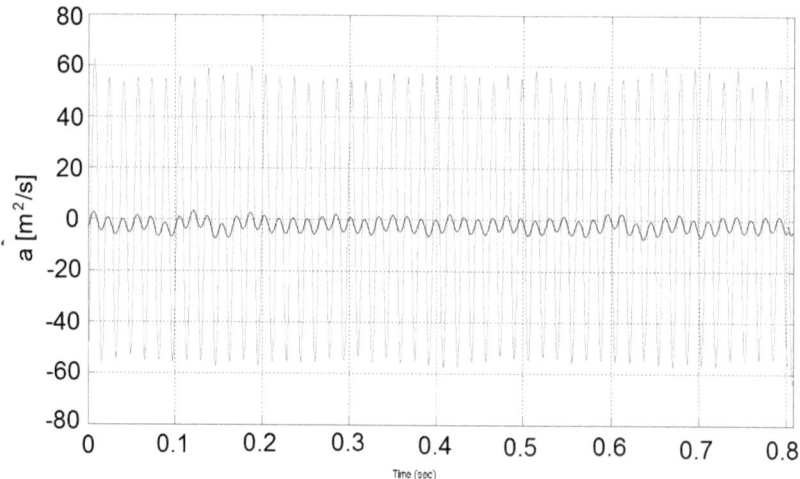

Fig. 4.25 The time waveform of the acceleration of the directional valve spool vibrations for the excitation frequency of 60 Hz. Blue color—solution after estimation of parameters of model 4 (4.33), gray color—results of experimental measurements

In addition, the ratio of the acceleration amplitudes obtained from the mathematical models (a_e) and the experiment (a_p) can be used to evaluate the models of the vibrating spool motion. Comparisons of the ratio of amplitudes for different equivalent stiffnesses of springs and oils of different viscosities are shown in Figs. 4.26, 4.27, 4.28 and 4.29.

Model 2 (Eq. 4.27), after parameterization and estimation of the constant coefficients C_1 and C_2, describes the spool vibrations in the directional valve sleeve much more precisely than model 1 (Eq. 4.17), model 3 (Eq. 4.32) and model 4 (4.33) over the entire frequency range, because the ratio of vibration amplitudes is closest to 1. Also, the value of the objective function shown in Table 4.10 is the smallest for model 2.

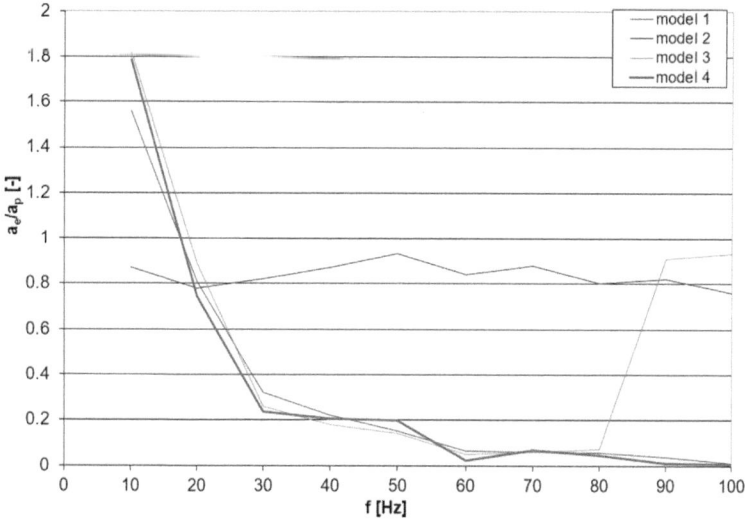

Fig. 4.26 Comparison of the value of the amplitude acceleration of the spool vibrations obtained from the experiment (a_p) and from the models (a_e) for the equivalent spring stiffness of 1632 N/m and HL68 oil

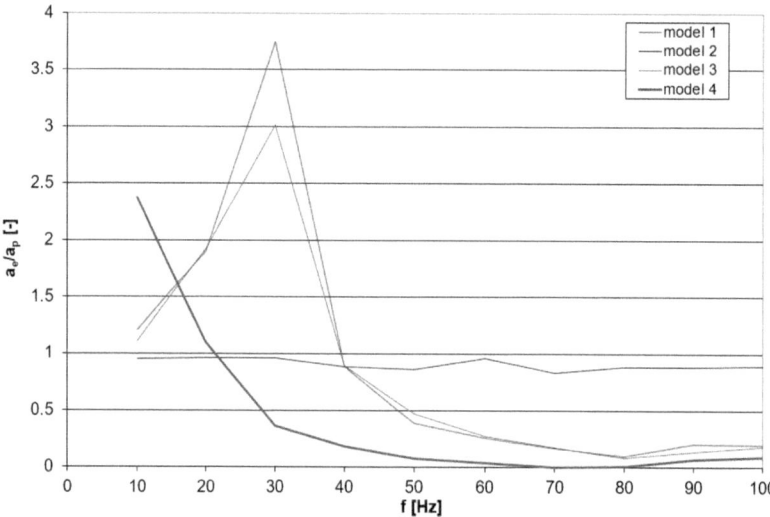

Fig. 4.27 Comparison of the value of the amplitude of the spool vibration acceleration obtained from the experiment (a_p) and from the models (a_e) for the equivalent spring stiffness of 5846 N/m and HL68 oil

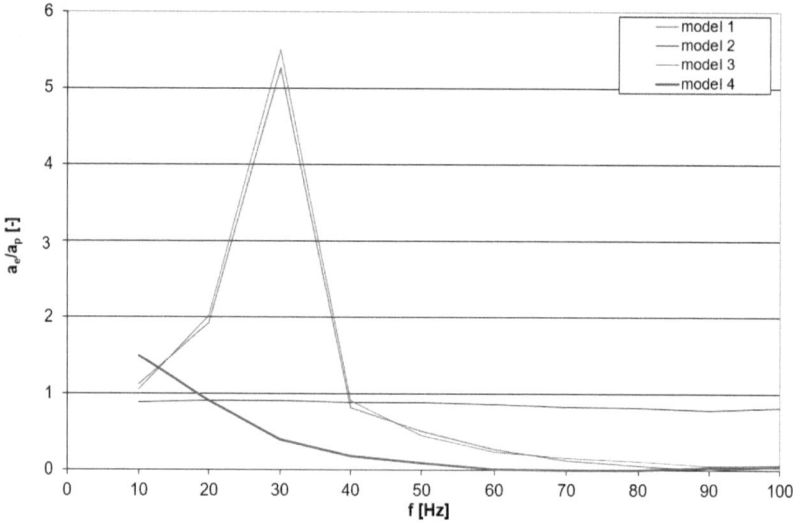

Fig. 4.28 Comparison of the value of the amplitude of the spool vibration acceleration obtained from the experiment (a_p) and from the models (a_e) for the equivalent spring stiffness of 5846 N/m and AzollaZS22 oil

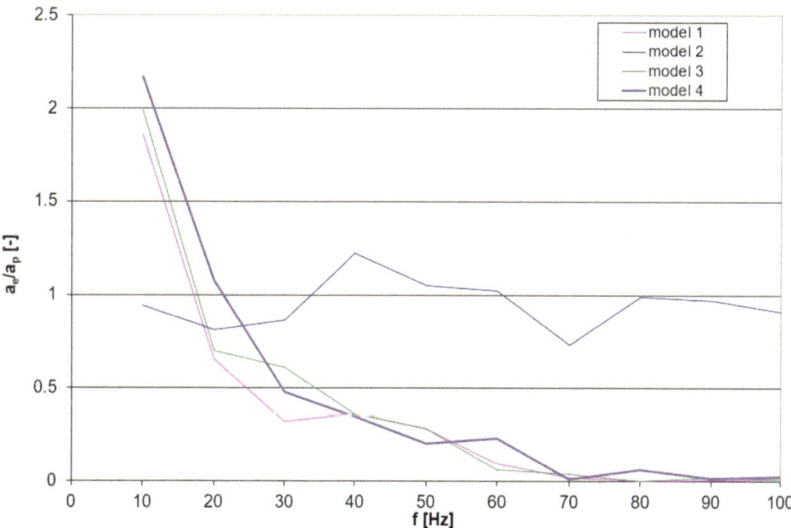

Fig. 4.29 Comparison of the value of the amplitude acceleration of the spool vibrations obtained from the experiment and from the models for the equivalent spring stiffness of 1632 N/m and AzollaZS22 oil

References

1. Chen, G. (Ed.). (2014) *Handbook of friction-vibration interactions*. Woodhead Publishing in Mechanical Engineering, Elsevier, Woodhead Publishing. ISBN 978-0-85709-458-2.
2. Hutchings, I., & Shipway, P. (2017) *Tribology: Friction and wear of engineering materials* (2nd ed.). Elsevier. ISBN 978-0-08-100910-9.
3. Gohar, R., & Rahnejat, H. (2018). *Fundamentals of tribology* (3rd ed.). World Scientific Publishing Company. ISBN 978-1-78634-519-6.
4. Khalil, M. (2018). *Hydraulic components volume A: Hydraulic sealing elements*. Amazon. ISBN 978-0-9977634-9-2.
5. Hol, J., Wiebenga, J. H., Hörning, M., Dietrich, F., & Dane, C. (2016). Advanced friction simulation of standardized friction tests. A numerical and experimental demonstrator. *Journal of Physics: Conference Series, 734*, 032092.
6. George, E., & Totten, E. (2018). *ASM handbook*. ASM International: Materials Park. ISBN 1-62708-141-0.
7. Carpinlioglu, M. O., & Gundogdu, M. Y. (2001). A critical review on pulsatile pipe flow studies directing towards future research topics. *Flow Measurement and Instrumentation, 12*, 163–174.
8. Bordovsky, P., & Murrenhoff, H. (2016) Investigation of steady-state flow forces in spool valves of different geometries and at different oil temperatures with the help of measurements and CFD simulations. In *Proceedings of the BATH/ASME 2016 Symposium on Fluid Power and Motion Control* (p. V001T01A013). American Society of Mechanical Engineers.
9. Stosiak, M., Karpenko, M., Deptuła, A., Urbanowicz, K., Skačkauskas, P., Deptuła, A. M., Danilevičius, A., Šukevičius, Š, & Łapka, M. (2023). Research of vibration effects on a

hydraulic valve in the pressure pulsation spectrum analysis. *Journal of Marine Science and Engineering, 11*, 301. https://doi.org/10.3390/jmse11020301

10. Stosiak, M., Karpenko, M., Prentkovskis, O., Deptuła, A., & Skačkauskas, P. (2023) Research of vibrations effect on hydraulic valves in military vehicles. *Defence Technology*. ISSN 2214-9147. https://doi.org/10.1016/j.dt.2023.03.023

11. Wen, S., & Huang, P. (2017). *Principles of tribology*. Wiley. 978-1-119-21489-2.

12. Heipl, O., & Murrenhoff, H. (2015). Friction of hydraulic rod seals at high velocities. *Tribology International, 85*, 66–73.

13. Khalil, M. (2022) Hydraulic systems volume 5 safety and maintenance. CompuDraulic LLC. ISBN 978-0-9977816-5-6.

14. Chatfield, C. (2017). *Statistics for technology: A course in applied statistics* (3rd ed.). Routledge. ISBN 978-0-203-73846-7.

Reduction the Impact of Vibration on a Hydraulic Drive Components—Valves and Pipelines

5

This chapter focuses on limiting the impact of external mechanical vibration on a hydraulic valve. Two approaches are considered to solve the problem of vibration being transferred to a valve and its control element (e.g., a spool). The first approach is to analyze the possibility of limiting vibration on the valve body by flexible mounting it on a vibrating substrate, i.e., a material with known stiffness and damping characteristics is placed between the valve body and the vibrating base. The second approach is the analysis of the effect of using spring-based damping elements inside the valve to reduce vibrations of the valve control, while the body is rigidly connected to a vibrating base. Additionally in this chapter presented a way of reducing vibration in hydraulic systems by appliying biomimetic approach—in example on hydraulic pipepelines.

Considering the reduction of the transmitted amplitudes and vibration frequency range, the placement of a specific material with appropriate elastic-and-dissipative properties can reduce the vibration amplitudes of the valve control elements. It is worth noting that the influence of mechanical vibration on valves and other hydraulic components, especially its consequences, are addressed in the normative solutions of the European Union.

Although the problem of vibrations in machines and devices is known and has been extensively analyzed, manufacturers of hydraulic components (e.g., valves) rarely provide information in the data sheets regarding the requirements for the product's resistance to vibration. However, occasionally a technical sheet contains information on the maximum values of mechanical vibrations that can affect the product, such as the hydraulic valve. The latest generation of high-speed proportional hydraulic directional valve Parker-Hannifin D1FP provides such information, where the datasheet claims that the maximum permissible value of vibration acceleration that may act on the directional valve is 250 m/s^2 [1]. Special standards define the method of testing the vibration resistance valves, e.g.,

© The Author(s), under exclusive license to Springer Nature Switzerland AG 2024   101
M. Stosiak and M. Karpenko, *Dynamics of Machines and Hydraulic Systems*, Synthesis Lectures on Mechanical Engineering, https://doi.org/10.1007/978-3-031-55525-1_5

EN 60068-2-6 [2], although other standards also describe the methods and conditions for testing. The standard PN-EN 60068-2-57:2013-12 [3] applies to the test method for components, devices, and other electrotechnical products, including electrically controlled hydraulic valves, which may be subjected to short-term pulsating or oscillating forces caused, for example, by seismic phenomena, explosion or vibration of the machines on which they are installed. The device is excited by several sinusoidal beats of constant frequencies. The standard PN-EN 60068-2-6 [4] provides a method for testing components, devices, and other products that may be exposed to harmonic vibrations during transport or operation, mainly produced by rotating, pulsating, or oscillating masses. Forces of this type are characteristic of ships, planes, ground vehicles, spaceships, etc. The tests are aimed at revealing the critical frequency, i.e., the frequency at which a product malfunctions or deteriorates due to vibration or mechanical resonances occurring (e.g., in tank covers, machine support frame, hydraulic valve control). The tests are carried out in a frequency range of 5–3000 Hz. Some standards are more general, such as PN-EN ISO 4413 [5], which provides requirements for the assembly of hydraulic components (including pumps, filters, and valves), recommending that the effects of gravity and vibration on hydraulic system components (including valves) should be considered.

Vibration resistance standards define the acceptable level of external mechanical vibration that may adversely affect a machine, device, or element. The permissible levels of vibration that may occur depend on the classes of machines or devices. To ensure the vibration resistance of measuring instruments and measuring apparatus, the precision industry adopted a maximum acceleration of approx. 0.981 m/s^2 [6] as a standard. This value is based on experimental studies, which state that such vibrations do not adversely affect the operation of these devices. However, rotating machines are allowed to undergo external mechanical vibrations of below 9.81 m/s^2, without their operation being affected.

Vibrations of hydraulic valve control elements are accompanied by noise with spectrum components in low and other frequencies. The noise level is an important criterion for assessing the quality of machinery and equipment. In terms of noise protection, the European Union Machinery Directive 2006/42/EC [7] established, inter alia, the following requirements: the machinery must be designed and constructed in such a way that the risks of emitted noise are reduced to the lowest level, in line with technical progress and considering available means of reducing noise, particularly at its source. This can be confirmed by the requirement set out in Directive 2005/88/EC [8] regarding the permissible sound power level of machines emitting noise to the environment.

In addition, since 15th February 2006, a new Directive 2007/30/EC [9] on protection against noise has been in force in all EU countries. It lowers the permissible maximum noise value at the operator's workplace, determined by averaging the values from 8 h exposure to the level of 80 dB(A) (previously 85 dB(A)).

5.1   The Use of Elastic Support to Reduce Vibrations on the Directional Valve Body

In this section, theoretical considerations are carried out on the vibration of the proportional valve body, which is mounted, with specially designed brackets that contain spring packs with known characteristics.

A vibration isolator, consisting of two springs connected in parallel, was used to reduce the vibrations of the hydraulic valve. The valve was mounted on a table holder and supported with springs on one side—Fig. 5.1. The table was vibrating. A simplified dynamic model of the valve mounted in a vibrating holder is shown in Fig. 5.2. In this case, the valve was treated as a single-mass system supported by springs on one side.

The designed holder allows the valve to be fastened and supported by springs on one side only (p. 6, Fig. 5.1a) with the equivalent stiffness c_z, and on the other side, it rests against a non-flexible bumper. The valve slides over the base of the holder (p. 2, Fig. 5.1a) rubbing against it according to the dry friction model. The designed holder has catches (p. 4, Fig. 5.1a) that prevent the valve from falling out when vibration acts on the valve. A clearance of l_0 is given between the valve body and the securing catches. When the clearance is removed, the valve body rubs against the securing catches according to the dry friction model.

The mathematical description containing the sum of the forces acting on the valve as a single-mass system is represented by the following equation:

$$m_2 \cdot \ddot{x}_{ko} + H(\dot{x}_{ko}) \cdot c_z \cdot (x_{0s} + x_{ko} - w) +$$
$$+ H(\dot{x}_{ko}) \cdot m_2 \cdot \mu_2 \cdot g \cdot (1 - H(l_0 - |x_{ko} - w|)) \cdot \text{sgn}(\dot{x}_{ko} - \dot{w}) + \qquad (5.1)$$
$$+ H(-\dot{x}_{ko}) \cdot m_2 \cdot \ddot{w} + \text{sgn}(\dot{x}_{ko} - \dot{w}) \cdot m_2 \cdot \mu_i \cdot g = 0$$

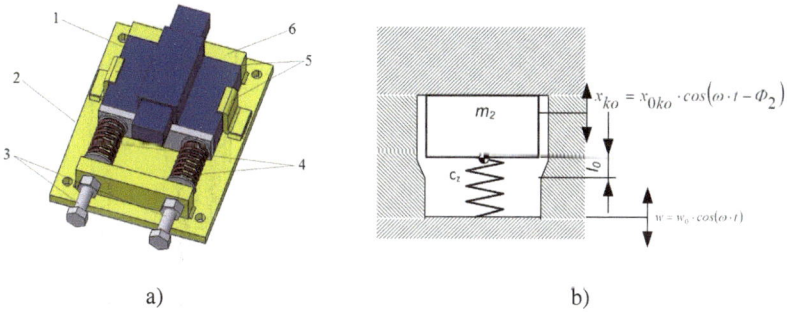

a)   b)

Fig. 5.1 Schemes for research № 1: **a** holder I of the valve: 1—hydraulic valve (e.g., directional valve), 2—holder base, 3—spring pre-deflection screws, 4—springs, 5—securing catches, 6—bumper; **b** model of a single-mass system in a holder with one-sided spring support

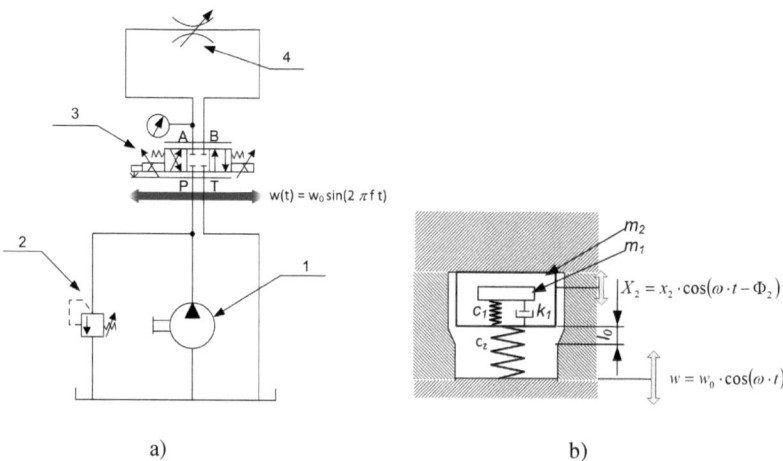

Fig. 5.2 Schemes for research № 2: **a** hydraulic diagram of the tested element's system: 1—feeding pump, 2—maximum valve, 3—tested element (proportional valve 4WRE 6 E08-12/24Z4/M), 4—adjustable throttle valve; **b** model of a two-mass system in a holder with one-sided spring support

where, m_2—mass of the valve; c_z—equivalent stiffness of the springs supporting the valve; x_{0s}—initial deflection of the springs; μ_2—coefficient of dry friction valve body against the protective holders; μ_i—coefficient of dry friction between the valve body against the holder base; g—gravitational acceleration; w—kinematic excitation; H—Heaviside function.

The hydraulic system that presents the operating directional valve under tests in a symbolic notation is in accordance with ISO 1219-1:2016 [10] in Fig. 5.2a. For a more detailed analysis of the behavior of the dynamic valve and the spool placed in the body, a dual-mass model can be used, in which mass m1 is the spool and mass m_2 is the valve body. The valve is supported on one side by springs, and on the other side, it rests against a non-flexible holder bumper—Fig. 5.2b.

The mathematical description of a vibrating directional valve working in a hydraulic system consists of a series of ordinary differential equation:

$$
\begin{cases}
m_1 \cdot \ddot{x}_{su} + \pi \cdot d_t \cdot \dfrac{l}{b} \cdot \mu \cdot (\dot{x}_{su} - \dot{x}_{ko}) + 0,72 \cdot \dfrac{1}{\sqrt{z}} \cdot 4 \cdot \dfrac{1}{2} \cdot s_s \cdot \dfrac{(x_{su} - x_p)}{r} \cdot (p_1 - p_2) \\[2mm]
+ c_{sz} \cdot (x_{su} - x_{ko}) = F_M \\[3mm]
Q_p - 4 \cdot \dfrac{1}{2} \cdot s_s \cdot \dfrac{(x_{su} - x_p)^2}{x_m} \cdot 0,75 \cdot \sqrt{\dfrac{2}{\rho} \cdot (p_1 - p_2)} - a_{p1} \cdot p_1 - c_{k1} \cdot \dfrac{dp_1}{dt} = 0 \\[3mm]
Q_p - a_{p1} \cdot p_1 - c_{k1} \cdot \dfrac{dp_1}{dt} - c_{k2} \cdot \dfrac{dp_2}{dt} - C_{q1} \cdot A_a \cdot \sqrt{\dfrac{2 \cdot p_2}{\rho}} = 0 \\[3mm]
m_2 \cdot \ddot{x}_{ko} + c_{sz} \cdot (x_{ko} - x_{su}) + k_{s1} \cdot (\dot{x}_{ko} - \dot{x}_{su}) + H(\dot{x}_{ko}) \cdot c_z \cdot (x_{0s} + x_{ko} - w) + \\[1mm]
+ H(-\dot{x}_{ko}) \cdot m_2 \cdot \ddot{w} + + H(\dot{x}_{ko}) \cdot m_2 \cdot \mu_2 \cdot g \cdot (1 - H(l_0 - |x_{ko} - w|)) \cdot sgn(\dot{x}_{ko} - \dot{w}) + \\[1mm]
+ sgn(\dot{x}_{ko} - \dot{w}) \cdot m_2 \cdot \mu_i \cdot g = 0
\end{cases}
$$

$$(5.2)$$

The first equations of the model (5.2) are in accordance with D'Alembert's principle, where the equilibrium of forces acting on a spool of mass m1 is excited to vibrate by the oscillating motion of the directional valve body. The spool is connected to the directional valve body via centering springs and damping in the gap of the spool pair. The next two Eqs. show the balance of the flow rate in the hydraulic system in which the tested valve operates. It is assumed that the maximum valve remains unopened, hence, the flow through this valve is not included in the second and third equations. The last equation of the model (5.2) is the sum of forces acting on the directional valve body while taking into account external mechanical vibrations (kinematic excitation) and interactions of the directional valve body with the simulator table holder. The presented equations are simplified so that the mathematical description is not too complicated, and at the same time, the results are reliable:

- The working fluid's properties are unchanged;
- Coulomb friction is omitted in the spool pair;
- Coulomb friction occurs during the relative motion of the body and the holder;
- Coulomb friction occurs after the clearance is removed during the relative movement of the body and securing catches;
- Springs that isolate body vibrations and springs centering the spool in the valve body have linear characteristics and are described by stiffness c;
- There are no wave phenomena in the hydraulic system;
- These assumptions can be utilized so the system can be described by ordinary differential equations.

A list of the main designations of the model equation parameters is included in Table 5.1.

In order to obtain a numerical solution to the model (5.2), the coefficients are parameterized. In this way, the values of the acceleration amplitude of the directional valve body

Table 5.1 List of important designations

Symbol	Parameter	Dimension in the SI
a_{p1}	Leakage rate	$[m^4 s/kg]$
A_a	Area of the throttle valve slot	$[m^2]$
c_1	Equivalent stiffness of springs centering the spool	$[N/m]$
c_z	Equivalent stiffness of the fixing springs in the holder	$[N/m]$
c_{k1}	Capacitance	$[m^5/N]$
C_{q1}	Flow rate through the throttle valve	$[-]$
d_t	piston diameter	$[m]$
f	Frequency	$[Hz]$
g	Gravitational acceleration	$[m/s^2]$
h	Thickness of the slot of the spool-sleeve pair	$[m]$
H	Heaviside function	$[-]$
k_{s1}, k_2	Damping in the spool-sleeve pair, Body-mounting bracket pair, respectively	$[N\ s/m]$
l	Piston length	$[m]$
l_0	The gap between the valve body and the safety brackets	$[m]$
m_1	The weight of the piston spool and 1/3 the weight of the spring	$[kg]$
m_2	Mass of the directional valve body	$[kg]$
p_1	Pressure upstream of the directional valve	$[Pa]$
p_2	Pressure downstream the directional valve	$[Pa]$
p_z	Pressure in the sink line	$[Pa]$
s_s	Maximum width of a gap	$[m]$
t	Time	$[s]$
w	Excitation vibration amplitude	$[m]$
Q_p	Theoretical pump capacity	$[m^3/s]$
x_m	Slot length	$[m]$
x_p	Spool's and the body's edges relatively shift	$[m]$
X_1	Displacement of the piston spool	$[m]$
X_2	Displacement of the directional valve body	$[m]$
μ_2	The coefficient of friction between the valve body and the safety catches	$[-]$
μ_i	Coefficient of friction between the valve body and the base of the holder	$[-]$
ρ	Working fluid density	$[kg/m^3]$
ω	Circular frequency	$[rad/s]$

are obtained for the given values of vibrations of the simulator table. In the numerical calculations, the following values of the coefficients are adopted: $m_1 = 0.0344$ kg, $m_2 = 4.5$
kg, $1 = 36.6 \times 10^{-3}$ m, $d_t = 12 \times 10^{-3}$ m, $h = 1.5 \times 10^{-7}$ m, $\mu = 0.22$ N $\times$ s/m, $\rho = 900$
kg/m^3, $\xi = 1.78$, $c_1 = 4884$ N/m, $l_0 = 0.2 \times 10^{-3}$ m, $\mu_2 = 0.1$, $\mu_i = 0.12$, $Q_p = 1 \times 10^{-4}$
m^3/s, $a_{p1} = 2.5 \times 10^{-11}$ m^{4}s/kg $A_a = 1.5 \times 10^{-6}$ m^2, $p_z \approx 0$ MPa, $c_{k1} = 0.62 \times 10^{-12}$ m^5/
N, $C_{q1} = 0.6$. The external kinematic excitation is described by a harmonic function of
the form $w(t) = w_{0(f)}\sin(2\pi\,f t)$ where $w_{0(f)}$—excitation amplitude [m] corresponding
to the excitation frequency, f—kinematic excitation frequency [Hz]. For the frequency f
$= 10$ Hz, $w_{0(10)} = 3.76 \times 10^{-3}$ m, for f $= 15$ Hz, $w_{0(15)} = 2.25 \times 10^{-3}$ m, f $= 20$ Hz,
$w_{0\,(20)} = 1.33 \times 10^{-3}$ m, for f $= 25$ Hz, $w_{0(25)} = 0.84 \times 10^{-3}$ m, for f $= 30$ Hz, $w_{0(30)} =$
0.48×10^{-3} m, for f $= 35$ Hz, $w_{0(35)} = 0.41 \times 10^{-3}$ m, for f $= 40$ Hz, $w_{0(40)} = 0.37 \times$
10^{-3} m, for f $= 45$ Hz, $w_{0(45)} = 0.27 \times 10^{-3}$ m, for f $= 50$ Hz, $w_{0(50)} = 0.21 \times 10^{-3}$ m,
for f $= 55$ Hz, $w_{0(55)} = 0.15 \times 10^{-3}$ m, for t $= 60$ Hz, $w_{0(60)} = 0.052 \times 10^{-3}$ m.

Figure 5.3 shows the value of the ratio between the amplitude of the vibration acceleration of the directional valve body a_2 and the amplitude of the vibration acceleration of
the simulator table a0.

If we assume that the vibration isolation is effective for the ratio of vibration acceleration amplitudes $a_2/a_0 < 1$, then it is examined for the frequency of the external mechanical
vibrations ≥ 30 Hz (Fig. 5.3). For lower values of excitation frequency, an amplification
of the directional valve body vibration amplitude is observed.

To increase the range of effective vibration isolation, the directional valve is considered
when supported by springs on both sides, as shown in Fig. 5.4.

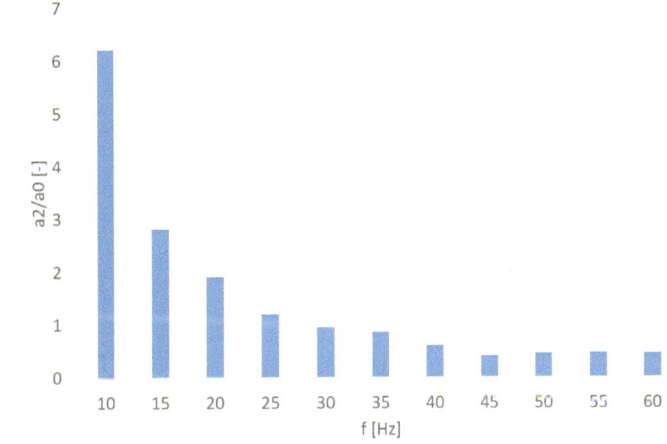

Fig. 5.3 Amplitude of vibration acceleration of the proportional valve body a_2 compared to the
amplitude of excitation vibration acceleration a_0 for f from 10 to 60 Hz. Model described by Eq.
(5.2)

Fig. 5.4 Valve holder II:
1—hydraulic valve (directional
valve), 2—holder base,
3—spring pre-deflection
screws, 4—springs,
5—securing catches

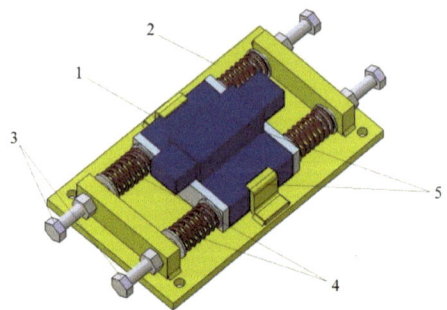

The holder is designed to allow the installed valve to be supported by springs on
both sides (p. 4, Fig. 5.4) of equivalent stiffness—c_z. The valve slides over the base of
the holder (p. 2, Fig. 5.4) rubbing against it according to the dry friction model. The
designed holder has catches (p. 5, Fig. 5.4) that prevent the valve from falling out when
vibration acts on the valve. A clearance of l0 is provided between the valve body and the
securing catches. When the clearance is removed, the valve body rubs against the securing
catches according to the dry friction model. The symbol of the hydraulic system in which
the directional valve operates is shown in Fig. 5.2a.

The dual-mass model can be used to analyze the dynamic behavior of the valve (body
and spool). The mathematical model of the directional valve excited to vibrate by external mechanical oscillations is given in the form of a series of four ordinary differential
equations:

$$
\begin{cases}
m_1 \cdot \ddot{x}_{su} + \pi \cdot d_t \cdot \dfrac{l}{h} \cdot \mu \cdot (\dot{x}_{su} - \dot{x}_{ko}) \\[2mm]
+ 0,72 \cdot \dfrac{1}{\sqrt{\xi}} \cdot 2 \cdot s_s \cdot \dfrac{(x_{su} - x_p)^2}{x_m} \cdot (p_1 - p_2) + c_{sz} \cdot (x_{su} - x_{ko}) = F_M \\[2mm]
Q_p - 1,5 \cdot s_s \cdot \dfrac{(x_{su} - x_p)^2}{x_m} \cdot \sqrt{\dfrac{2}{\rho} \cdot (p_1 - p_2)} - a_{p1} \cdot p_1 - c_{k1} \cdot \dot{p}_1 = 0 \\[2mm]
Q_p - a_{p1} \cdot p_1 - c_{k1} \cdot \dot{p}_1 - c_{k2} \cdot \dot{p}_2 - C_{q1} \cdot A_a \cdot \sqrt{\dfrac{2 \cdot p_2}{\rho}} = 0 \\[2mm]
m_2 \cdot \ddot{x}_{ko} + c_{sz} \cdot (x_{ko} - x_{su}) + k_{s1} \cdot (\dot{x}_{ko} - \dot{x}_{su}) + c_z \cdot (x_{0s} + x_{ko} - w) + \\[1mm]
+ m_2 \cdot \mu_2 \cdot g \cdot (1 - H(l_0 - |x_{ko} - w|)) \cdot \mathrm{sgn}(\dot{x}_{ko} - \dot{w}) \\[1mm]
+ \mathrm{sgn}(\dot{x}_{ko} - \dot{w}) \cdot m_2 \cdot \mu_i \cdot g = 0
\end{cases}
\tag{5.3}
$$

In the series of Eq. (5.3), the first three are the same as for the model (5.2), because
the physical model of the directional valve's spool-body pair and the configuration of
the hydraulic system remains unchanged. The last Eq. changes because the design of the
simulator table holder is different. It is assumed that as a result of the relative movement

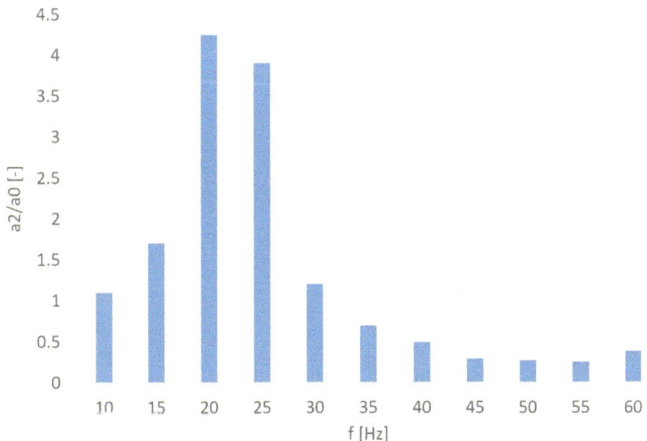

Fig. 5.5 Amplitude of vibration acceleration of the proportional valve body a_2 compared to the amplitude of excitation vibration acceleration a_0 for f from 10 to 60 Hz. Model described by Eq. (5.3)

of the directional valve body and the holder, the clearance l0 between the securing catches and the directional valve body is eliminated. Hence, dry friction occurs. The fourth Eq. can be modified by taking into account the different models of vibration isolators. The main notations appearing in the model's Eq. (5.3) are summarized in Table 5.1.

After adopting numerical values for the parameters of the model's Eq. (5.3), it can be solved numerically. The ratio between the value of the amplitude of vibration acceleration of the valve body a_2 and the value of the amplitude of vibration acceleration of the forcing a_0 is adopted, and this ratio is called the "transfer function"—Fig. 5.5. The numerical values of the model parameters (5.3) are the same as for the model (5.2). Also, the kinematic excitation is not changed.

Figure 5.5 shows a clear mechanical resonance in the form of amplification of the acceleration amplitude of the directional vibration of the valve body for a frequency of approx. 20 Hz. When the mass of the directional valve is equal to 4.5 kg and the equivalent stiffness of the springs in the holder, then the natural frequency is approx. 22 Hz. For the external excitation frequency range of 10–30 Hz, ineffective vibration isolation is observed because the value of the ratio a_2/a_0 is greater than 1.

It is necessary to examine a type model of vibration isolator that allows for the extension of the range of effective vibration isolation. The "black box" method (Fig. 5.6) has been used to study the effects of the vibration isolator characteristics on the value of the "transfer function".

The previous considerations allow modification of the directional control valve vibration model, taking into account different characteristics of vibration isolators. Equation (5.3) can be written for a vibration isolator with linear stiffness and damping

Fig. 5.6 The "black box" method in the isolation of a valve against vibration

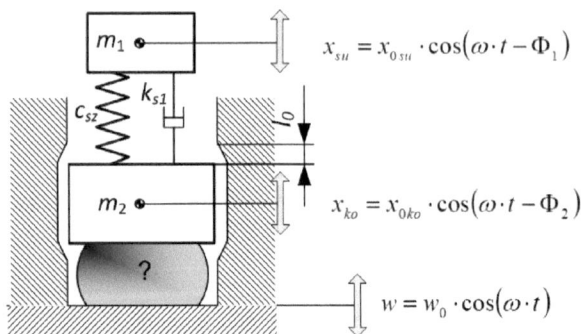

$$x_{su} = x_{0su} \cdot \cos(\omega \cdot t - \Phi_1)$$

$$x_{ko} = x_{0ko} \cdot \cos(\omega \cdot t - \Phi_2)$$

$$w = w_0 \cdot \cos(\omega \cdot t)$$

characteristics and the Kelvin-Voigt body model $c_z \cdot (x_{ko} - w) + k_2 \cdot (\dot{x}_{ko} - \dot{w})$ while assuming that the clearance l_0 is not eliminated. Hence, the fourth Eq. in the series (5.3) can be written as follows:

$$m_2 \cdot \ddot{x}_{ko} + c_{sz} \cdot (x_{ko} - x_{su}) + k_{s1} \cdot (\dot{x}_{ko} - \dot{x}_{su}) + c_z \cdot (x_{0s} + x_{ko} - w) + k_2 \cdot (\dot{x}_{ko} - \dot{w}) +$$
$$+ sgn(\dot{x}_{ko} - \dot{w}) \cdot m_2 \cdot \mu_i \cdot g = 0$$

$$(5.4)$$

In order to assess the impact of change towards the damping value in the spool pair, a constant equivalent stiffness of the vibration isolator springs is assumed, $c_z = 20{,}000$ N/m, and two damping values $k_2 = 50$ Ns/m and $k_2 = 250$ Ns/m. The stiffness change effect can be evaluated by altering the equivalent stiffness value at a fixed damping value. Thus, for the set damping value $k_2 = 50$ Ns/m, the equivalent stiffness is changed to 120,000 N/m. Figure 5.7 shows the values of the "transfer function" for different values of the parameters c_z and k_2.

The analysis shown in Figs. 5.5 and 5.7 indicates that a vibration isolator with a linear characteristic and different values of c_z and k_2 parameters can improve the effectiveness of the vibration isolation (by limiting the vibration amplitudes of the directional valve body and reducing the resonance zone). The effectiveness of vibration isolation can be increased by using non-linear vibration isolators.

For example, when a vibration isolator with a non-linear stiffness characteristic is implemented, the spring stiffness force is proportional to the square of the relative displacement and a linear damping characteristic. In other words, the utilization of a damping force proportional to the relative speed, and with the play l_0 not being canceled, the last Eq. (5.3) can be written as follows:

$$m_2 \cdot \ddot{x}_{ko} + c_{sz} \cdot (x_{ko} - x_{su}) + k_{s1} \cdot (\dot{x}_{ko} - \dot{x}_{su}) + c_z \cdot (x_{0s} + x_{ko} - w)^2 + k_2 \cdot (\dot{x}_{ko} - \dot{w}) +$$
$$+ sgn(\dot{x}_{ko} - \dot{w}) \cdot m_2 \cdot \mu_i \cdot g = 0$$

$$(5.5)$$

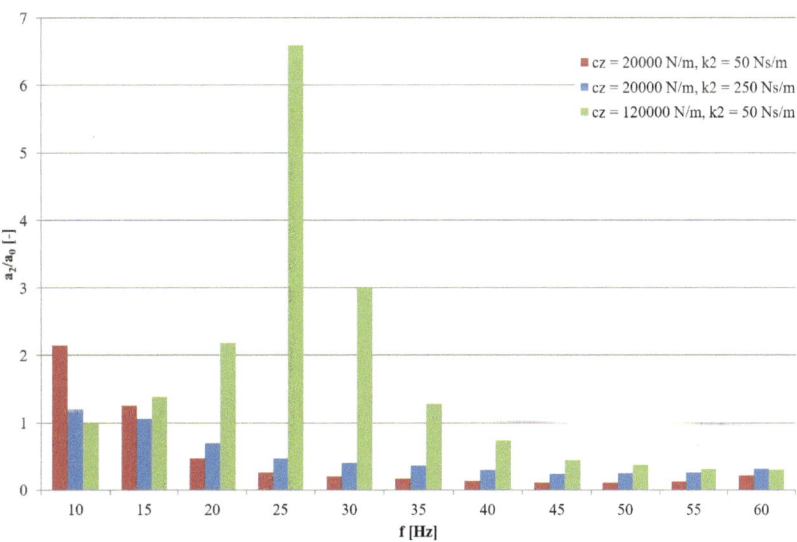

Fig. 5.7 Amplitude of vibration acceleration of the proportional valve body a_2 compared to the amplitude of excitation vibration acceleration a_0 for f from 10 to 60 Hz. Model modified by Eq. (5.4)

The effect of damping on the reduction of vibrations of the valve body can be assessed when a constant equivalent stiffness equal to $c_z = 20,000$ N/m and a change in the damping coefficient value $k_2 = 50$ or 250 Ns/m can be adopted. In this case, a numerical solution describing the amplitude of vibration acceleration of the directional valve body for the considered frequency of external vibration in the range of 10–60 Hz is achieved. The "transfer function" for selected excitation frequencies is shown in Fig. 5.8, for a constant value of equivalent stiffness and two damping values of the vibration isolator material.

The analysis shown in Fig. 5.8 displays that for a constant value of the equivalent stiffness, an increase in the damping value causes an increase in the "transfer function". However, the use of a vibration isolator with the proposed characteristics makes the vibration isolation effective over the entire range of analyzed frequencies.

In contrast to the aforementioned case, one can consider the effects of using a vibration isolator with linear stiffness characteristics ($c_z = 20,000$ N/m) and non-linear damping characteristics ($k_2 = 250$ Ns^2/m^2) of the form $c_z \cdot (x_{ko} - w) + k_2 \cdot (\dot{x}_{ko} - \dot{w})^2$. The resulting transfer function values are shown in Fig. 5.9.

In this case, the extension of the effective vibration isolation zone can be observed, which also includes the resonant frequency of the spool of a typical single-stage directional valve.

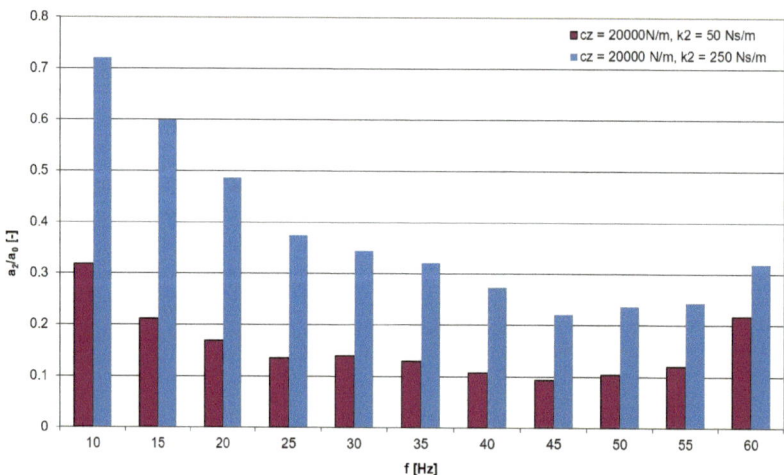

Fig. 5.8 Amplitude of vibration acceleration of the proportional valve body a_2 compared to the amplitude of excitation vibration acceleration a_0 for f from 10 to 60 Hz (transfer function). Model modified by Eq. (5.5)

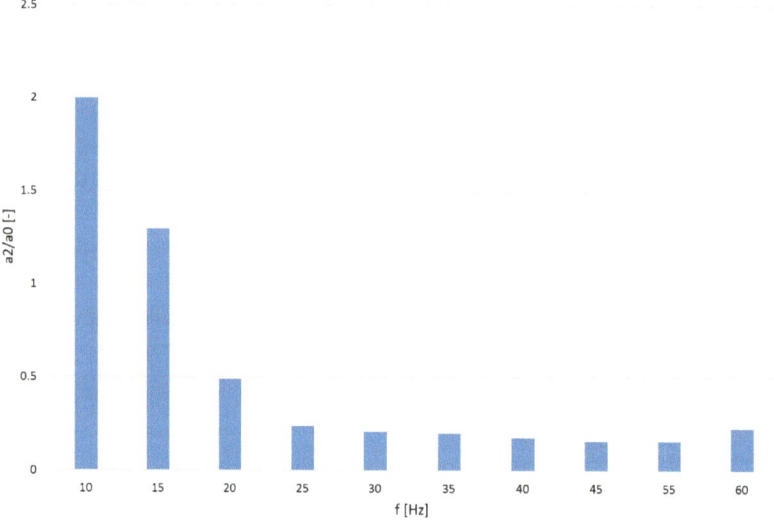

Fig. 5.9 Amplitude of vibration acceleration of the proportional valve body a_2 compared to the amplitude of excitation vibration acceleration a_0 for f from 10 to 60 Hz. A vibration isolator with linear stiffness characteristics and non-linear damping characteristics

5.2 Experimental Verification of the Possibility of Reducing the Vibration of a Hydraulic Valve Using Spring Packs and Elastomeric Elements

These analytical considerations and numerical solutions were verified experimentally to the extent possible from the point of view of implementation. Experimental tests consisted of subjecting a single-stage proportional directional valve with a spool structure Rexroth 4WRE 6 E08-12/24Z4/M to external mechanical vibration. The directional valve was installed in the vibrating holder of the simulator table, acting as a vibration generator. During the tests, two types of holders were used (special holder I and special holder II), enabling different ways of installing vibration isolators, Fig. 5.10 regarding—Figs. 5.1a and 5.4.

In the first series of tests, a proportional directional valve was placed in special holder I and supported by a set of two springs, on one side. The stiffness of each spring was 43,000 N/m. The springs formed a parallel system of connections, so their stiffness was 86,000 N/m, and the initial spring deflection was 2 mm. After the first series of tests, they were repeated with different springs, in which the stiffness was 22,000 N/m. Thus, in the second measurement series, the equivalent stiffness of the system of two parallel connected springs was 44,000 N/m. The results of the experimental tests in the form of a bar graph of the transfer function are shown in Fig. 5.11.

The results presented in Fig. 5.11 indicate that for a spring vibration isolator with equivalent stiffness $cz = 86,000$ N/m, two ranges were identified for the vibrating motion of the proportional directional valve: the range where resonance occurs, i.e., at approx. 25 Hz and the range of effective vibration isolation (except for 60 Hz). Hence, in this case, the amplitude of the acceleration of vibration of the directional valve decreased

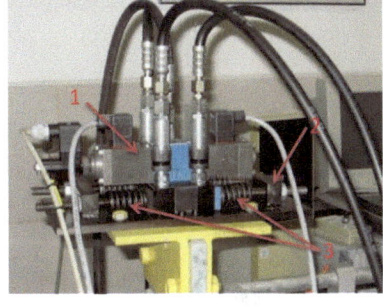

a)   b)

Fig. 5.10 Proportional valve during experimental tests mounted (1—hydraulic valve, 2—vibrating special holder, 3—set of springs) in: **a** special holder I and supported on one side by springs (set of two springs); **b** special holder II and supported by springs on both sides (set of fourth springs)

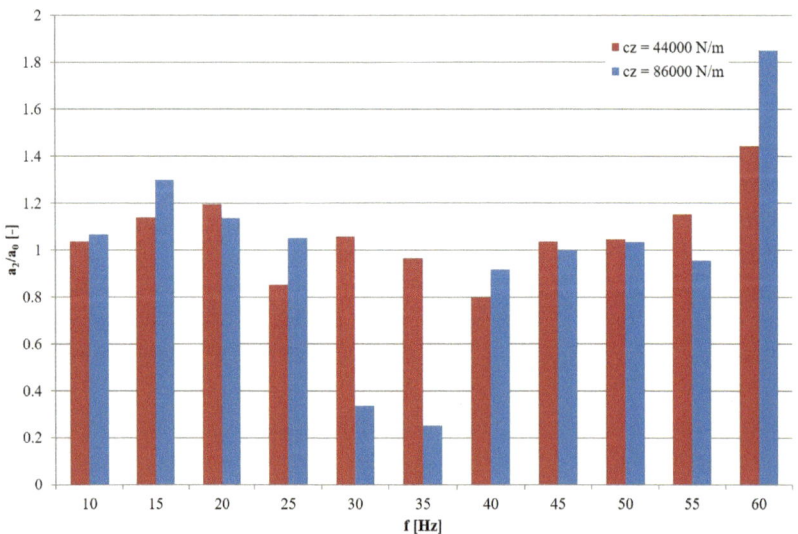

Fig. 5.11 Amplitude of vibration acceleration of the proportional valve body a_2 compared to the amplitude of excitation vibration acceleration a_0 for f from 10 to 60 Hz. A proportional directional valve placed in special holder I and supported by a set of two springs, on one side

slightly, and the amplitude of vibration acceleration for the resonant frequency increased significantly. An even greater increase in the amplitude of vibration acceleration was observed for a frequency of 60 Hz.

Therefore, experimental research was carried out on the effectiveness of reducing the vibration of the directional valve using spring vibration isolators that protect the housing of the directional valve on both sides. For this purpose, the directional valve was placed in special holder II (Fig. 5.4). The equivalent stiffness of such a vibration isolator was 86,000 N/m, and the initial deflection of the springs was set at 2 mm. The results are presented in Fig. 5.12.

The obtained results (Fig. 5.12) proved that the applied method of isolation of the directional valve vibration was more effective than one-sided support of the directional valve with a vibration insulator (Fig. 5.1). Therefore, the area of effective vibration isolation was extended and ranged from 25 to 60 Hz—Fig. 5.13.

These experimental results showed that the use of spring visor insulators positively affected the reduction of the vibration of the directional valve. However, such a solution showed that there were vibration frequencies at which the directional control valve vibration was amplified, and, in addition, the spring vibration isolator requires the necessary space for the application. This could be problematic under the conditions of the machine or equipment. The space occupied by the vibration isolation system could be minimized,

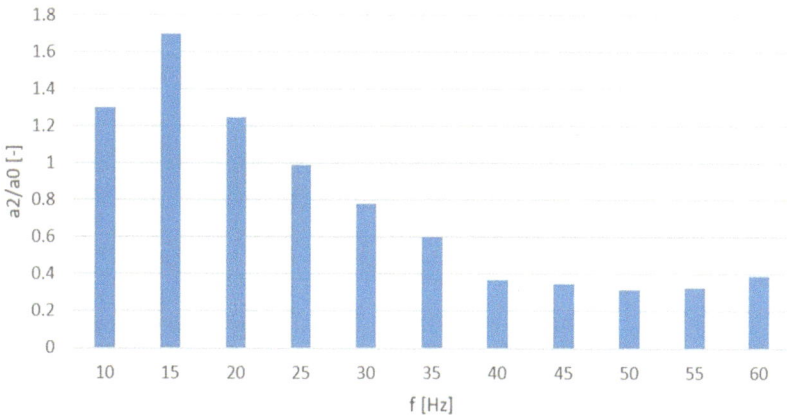

Fig. 5.12 Amplitude of vibration acceleration of the proportional valve body a_2 compared to the amplitude of excitation vibration acceleration a_0 for f from 10 to 60 Hz. A proportional directional valve placed in special holder II and supported by a set of four springs, on both sides

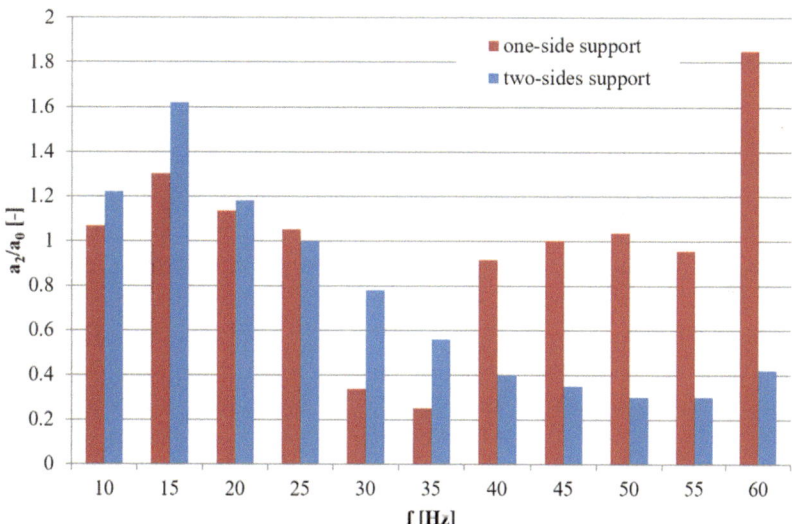

Fig. 5.13 Amplitude of vibration acceleration of the proportional valve body a_2 compared to the amplitude of excitation vibration acceleration a_0 for f from 10 to 60 Hz. Comparison of valve mounting method: single-sided and both sides

Fig. 5.14 The hydraulic directional valve mounted in the holder with flexible elements that isolate vibrations: 1—simulator table; 2—holder; 3—hydraulic directional valve; 4—a set of flexible elements that isolate vibrations; 5—accelerometer for measuring vibration acceleration of the directional valve body (answer); 6—accelerometer for measuring vibration accelerations of the simulator table (excitation)

and the effectiveness of vibration isolation further increased by using, for example, isolators made of elastomeric materials. In addition, the vibration acceleration amplitude could be restricted by making the vibration insulators from materials that dissipate energy.

Therefore, experimental studies were carried out using elastomeric elements as a vibration isolator material. During the experimental tests, the vibration acceleration of the simulator table (excitation) and the directional valve body (response) were measured—Fig. 5.14.

The experiments were carried out using accelerometers PCB–ICP, signal conditioner VibAmp PA16000D, four-channel digital oscilloscope Tektronix with specialized software provided by the manufacturer, and a personal computer for measurement data acquisition. Before the main tests, separate tests were conducted to identify the elasticity and dissipativity of the materials used as vibration insulators. Two types of samples were examined: cylindrical and cuboid samples. Identification tests were performed using instruments designed for experimental modal analysis [11, 12]. In the tests, a modal hammer with a built-in force sensor was used as the excitation, while the response of the system was recorded using an acceleration sensor. Measurement signals were recorded and processed by a specialized analyzer. The analysis of the obtained results was carried out using a PC with the appropriate software. The data was obtained on the elastic-dissipative properties of the samples and was later used for selecting vibration-isolating material.

A custom-built testing stand is shown in Fig. 5.15a. The main components of the test stand were: seismic mass weighing 49 kg; vibrating mass weighing 12.9 kg; a set of anti-vibration shims; control and measurement equipment with measurement circuits.

The measuring instruments included a modal hammer from PCB Piezoelectronics (type: SP 205) with a built-in force sensor, which allowed measurement of the amplitude of the applied excitation. In addition, it possessed a PCB accelerometer to measure

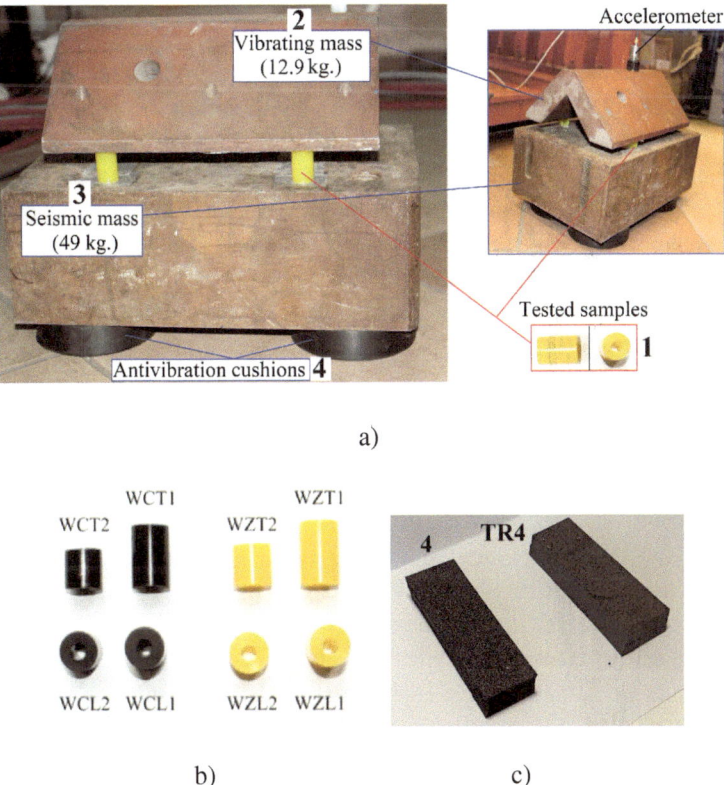

Fig. 5.15 Some views from vibration experiment test: **a** test stand: 1—tested samples; 2—vibrating mass; 3—seismic mass, 4—anti-vibration shims; **b** designations of cylindrical samples (with a top view); **c** designations of rectangle samples

the system's response to the applied modal hammer forcing, and a signal analyzer from HP (type: 35665A) to record sensor data and process it into transition characteristics.

The aim of the study was to determine the frequency characteristics for each type of sample. On their basis, the parameters of damping and dynamic stiffness were then determined for samples of different shapes and made of different materials. The tests were carried out in such a way that first, the test specimens were placed on a seismic mass, then a steel element was laid on top as a vibrating mass. All samples of the same type were arranged in the same, strictly defined manner. To determine the elasticity and dissipativity of cylindrical samples due to their small dimensions, four samples (of the same type, shape, and material) were used simultaneously during testing. Additionally, to ensure that all cylindrical specimens were uniformly loaded during dynamic testing, they were arranged so that they formed a square with a side of 14 cm. However, the cuboidal samples were arranged in pairs, parallel to each other, so that the vibrating mass oscillated in the vertical axis.

In the case of cylindrical elastomer shims, measurements were carried out on two types of orientations of the samples in relation to the vibrating mass. In the first case, the long axis of the sample (symmetry axis) was oriented along the vertical axis, i.e., parallel to the direction of motion of the vibrating mass. In the second case, the long axis of the samples was rotated by 90°, and lay horizontally, i.e., oriented perpendicularly to the direction of vibration. The samples and their designations used during the tests are shown in Fig. 5.15b. The letter W denotes a cylindrical sample, while the letter C or Z (second in order) denotes the type of material "Adipol 70 ShA" and "Ultraflex 64 ShA", respectively. The letters T and L relate to the direction of sample loading: L—longitudinally to the axis of symmetry of the sample; T—transverse to the axis of symmetry. The number 1 at the end of a designation means a sample length of 25 mm, while the number 2 means a sample length of 16 mm. For all cylindrical elastomers, the outer diameter was 16 mm and the inner diameter was 6 mm.

The tests were conducted for rectangular samples made of oil-resistant rubber, which were mounted in a holder to reduce the vibration on the directional valve body. All rectangular samples had the same length and width, i.e., 101 mm and 26 mm, respectively, while the thickness was 4, 12, and 15 mm. The nomenclature adopts numerical designations that define the thickness of the sample, while "TR" denotes a material with increased stiffness. Based on the identification tests, the characteristics of the transition were obtained from the waveforms of force and acceleration using a signal analyzer, in which the transition was a function of frequency. Examples of transition characteristics for the sample are marked with the symbol 12 in Fig. 5.16a and for WCT1 at Fig. 5.16b.

The experimental data is all for the determination of the time and frequency response of the vibrating mass and the tested vibration isolator subjected to impulse excitations.

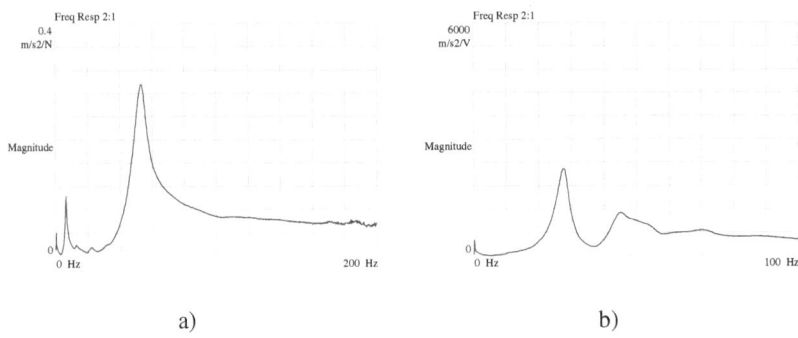

a)   b)

Fig. 5.16 Frequency characteristics of: **a** frequency characteristics for sample 12 made of oil-resistant rubber: f—vibration frequency of the vibrating mass, fr—natural frequency of the vibrating mass.; **b** WCT1 cylindrical sample, which consisted of material Adipol 70ShA

Then, using the half-power method [11], the modal damping was determined ξ_r for each type of vibration isolator according to Eq. 5.6:

$$\xi_r = \frac{\omega_2 - \omega_1}{2\omega_r}. \tag{5.6}$$

while the frequencies ω_1 and ω_2 were determined using the following equation:

$$|H(j\omega_1)| = |H(j\omega_2)| = \frac{|H(j\omega_r)|}{\sqrt{2}}, \tag{5.7}$$

where, $|H(j\omega_r)|$ is the amplitude value of the transition function corresponding to the natural frequency ω_r; $|H(j\omega_1)|$ and $|H(j\omega_2)|$ are the amplitudes of the transition function corresponding to the frequencies ω_1 and ω_2, lying on both sides of the natural frequency ω_r according to the condition (Eq. 5.7) $\omega_r = 2\pi f_r$.

Assuming that the tested real system was treated as a system with one degree of freedom with parallel-connected elastic-dissipative elements, the following equation was expressed:

$$h_r = \frac{k}{2m} \tag{5.8}$$

and:

$$h_r = \xi_r \cdot f_r \tag{5.9}$$

The comparison of Eqs. (5.8) and (5.9) side by side determined the damping k. However, the f_r value was determined from the experiment, which was related to the frequency response for pulsed excitation. Moreover, when $\omega_r = 2\pi f_r$ and the following equations was used:

$$\omega_r = \sqrt{\omega_0^2 - 4h_r^2} \tag{5.10}$$

$$\omega_0 = \sqrt{\frac{c}{m}} \tag{5.11}$$

The stiffness value of the material that isolated the vibrating mass can be determined. To describe the dynamic properties of materials used in vibration isolators, a simple dynamic model with one degree of freedom was employed—Fig. 5.17, which consisted of linear stiffness and damping connected in parallel. This was the Kelvin–Voigt model and mass m represents the vibrating mass.

Due to the arrangement of the samples during testing, they were treated as elements connected in parallel (same deflection arrow at the same time), where the stiffness and damping values of individual samples were determined from Eq. (5.12) for cylindrical samples and Eq. (5.13) for rectangular samples:

Fig. 5.17 Diagram of the dynamic model: c—stiffness parameter; k—damping parameter; m—vibrating mass

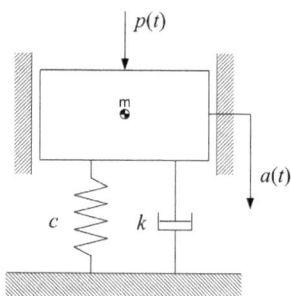

$$c = \tfrac{c_z}{4};\; k = \tfrac{k_z}{4}; \tag{5.12}$$

where c_z—equivalent stiffness for the tested set of samples, k_z—equivalent damping for the tested set of samples.

In the case of cuboid samples:

$$c = \tfrac{c_z}{2};\; k = \tfrac{k_z}{2}. \tag{5.13}$$

Tables 5.2 and 5.3 show the values of stiffness and damping for the considered types of samples.

The obtained stiffness and damping values were used to identify the dynamic properties of the materials to analyze the effectiveness of vibration reduction. The results inferred that the elasticity and dissipativity were significantly affected by the sample's orientation. When the long axis of the sample was oriented parallel to the direction of motion of the vibrating mass, then the tested elements had higher damping and stiffness than when the long axis was oriented perpendicularly to the direction of the load. Significantly lower damping and stiffness values were obtained for samples made of the material Ultraflex 64 ShA compared to the samples made of the material Adipol 70 ShA. However, for cuboid

Table 5.2 Damping parameters k and stiffness c of the cylindrical samples

Sample	Damping k, [kg/s]	Stiffness c, [kg/s^2]
WCL1	10.97	9.70×10^4
WCT1	7.26	4.30×10^4
WCL2	9.27	8.04×10^4
WCT2	8.47	6.49×10^4
WZL1	7.01	8.41×10^4
CTM1	3.55	2.79×10^4
WZL2	4.69	5.55×10^4
CTM2	4.43	4.30×10^4

Table 5.3 Damping parameters k and stiffness c of the rectangular samples

Sample	Damping k, [kg/s]	Stiffness c, [kg/s^2]
4	112.88	2.16×10^6
12	59.02	7.50×10^5
15	52.57	7.16×10^5
TR 4	149.96	4.08×10^6
TR 12	91.27	1.44×10^6
TR 15	66.11	1.16×10^6

samples, the values of damping and stiffness parameters decreased with increasing sample thickness.

In the experimental tests, the vibration isolators of the sample, whose dynamic properties were determined above, were used to reduce the vibration of the directional valve (Tables 5.2 and 5.3). Figure 5.18 show the positioning of vibration isolators (regarding Fig. 5.14). The effectiveness of vibration isolation was tested for the frequency range of 10–100 Hz.

In special holder II, the sets of the above-listed elastomer samples were installed on both sides. The samples with identical geometries and materials were used each time. The effectiveness of the directional valve vibration reduction was assessed using the transfer function, which was extrapolated from the ratio between the directional valve vibration acceleration amplitude and the simulator table vibration acceleration amplitude. The vibration isolation was effective when the value of the transfer function was below 1:

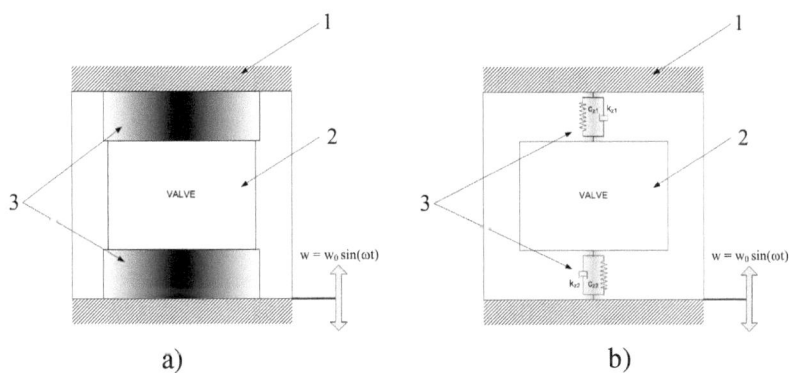

a)   b)

Fig. 5.18 The directional valve's flexible attachment to the substrate: **a** location of flexible elements; **b** dynamic model of flexible elements: 1—vibrating holder of the simulator table; 2—hydraulic directional valve; 3—set of flexible elements isolating vibration

$$\frac{a_2}{a_0} < 1, \tag{5.14}$$

where: a_2—amplitude of the directional valve body vibration acceleration, m/s^2, and a_0—vibration acceleration amplitude of the simulator table, m/s^2.

Five different vibration isolator samples consisting of different geometry and materials were used in the tests. Individual sets differed in the number of elastomeric elements they consisted of and in the direction of the load in relation to the long axis for the cylindrical samples. As a result, sets with different equivalent stiffness and different equivalent damping were obtained.

The considered sets of shims with different configurations had equivalent stiffness and equivalent damping. The values are given in Table 5.4.

The obtained results in the form of a cumulative graph show the ratio between the amplitudes of the vibration acceleration of the directional valve and the simulator table, depending on the excitation frequency (Fig. 5.19).

The results shown in Fig. 5.19 indicate that the appropriate selection of vibration-isolating material allows for effective vibration isolation in a wide range of vibration frequencies. Reduction of vibrations of the directional valve body means reducing the excitations acting on the directional valve spool through the stiffness forces of the centering springs and friction processes in the spool pair.

Table 5.4 Values of equivalent stiffness c_z and equivalent damping k_z for the considered sets of shims

Shims set number	Equivalent stiffness c_z, [N/m]	Equivalent damping k_z, [kg/s]
1	28.36×10^4	43.24
2	2.32×10^6	132.22
3	54.36×10^4	55.84
4	65.52×10^4	70.04
5	71.56×10^4	84.88

Fig. 5.19 The efficiency of isolation of directional valve vibration forced by ground vibration for different sets of elastic shims: a_2—body acceleration amplitude, a_0—simulator table acceleration amplitude

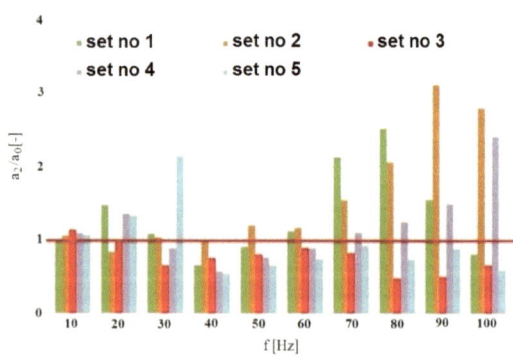

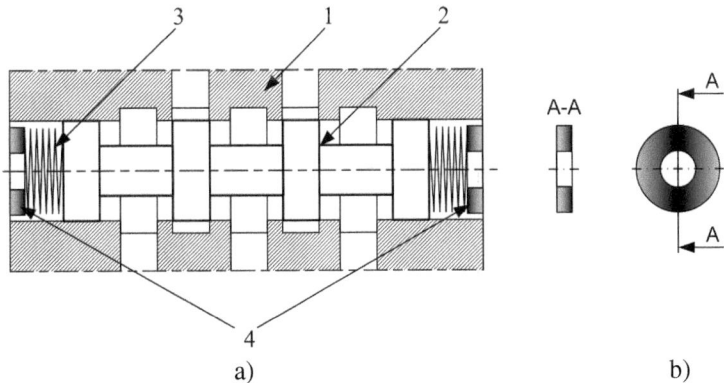

Fig. 5.20 Shims made of oil-resistant rubber installed inside the directional valve body: **a** location inside the directional valve body: 1—vibrating directional valve body; 2—directional valve spool; 3—centering springs; 4—flexible shims; **b** shape of a shim

Independently or in addition to the presented method of reducing the impact of external vibrations on the directional valve, the valve can be modified with by inserting custom-profile flexible shims (e.g., oil-resistant rubber) between the valve body and the centering springs—Fig. 5.20. Experimental testing was carried out to verify the possibility of using such a solution. The directional valve was mounted directly in the vibrating holder of the simulator table. The simulator table's vibration frequency ranged from 10 to 100 Hz.

The internal shims were designed and prepared so their shape reflected the specifics of the directional control valve design. The outer diameter was 26 mm, and the inner diameter was 22 mm, with a thickness of 4 mm.

The dynamic model of the spool as a single-mass system inside the directional valve body is shown in Fig. 5.21. The spool weighing m_1 was centered by springs of known stiffness, c_{s1} for each of the spring. In addition, damping k_{s1} on each side of the spool pair was considered.

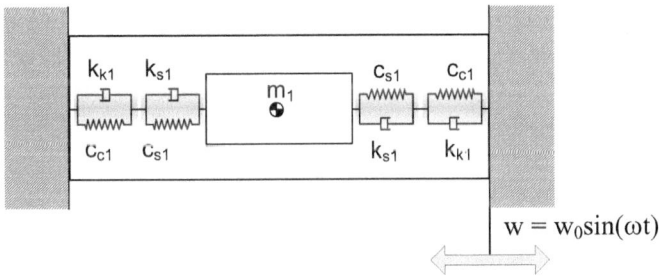

Fig. 5.21 A dynamic model of the spool in a vibrating directional valve body

Fig. 5.22 The ratio between the relative acceleration of the spool and the acceleration of the valve body vibration as a function of the body vibration frequency

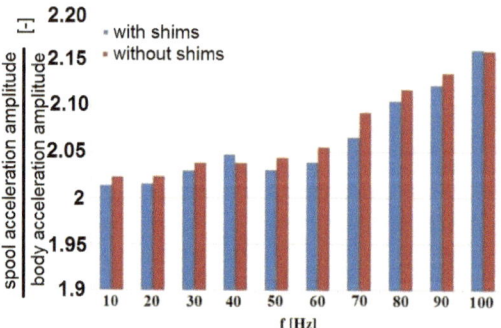

It was assumed that the shims had linear stiffness and damping characteristics that could be described by the Kelvin–Voigt body model. For each of the two shims, the stiffness was denoted by c_{c1} and the damping by k_{k1}.

In the tested directional valve, the stiffness of the centering spring was $c_{s1} = 2900$ N/m. The shim's stiffness was $c_{c1} = 1.41 \times 10^6$ N/m. When the stiffnesses were connected in series, the stiffness on each side of the spool was 2894 N/m. On both sides of the spool, there were elements (centering spring and shim) of equal stiffness, which were connected in parallel to each other, resulting in a stiffness of 5788 N/m. A similar procedure was followed for the determination of the equivalent damping, which eventually amounted to 17.32 kg/s. After inserting shims inside the directional control valve (between the centering springs and the valve body), experimental tests were carried out, and the results are shown in Fig. 5.22.

Due to the series connection of stiffnesses, an equivalent stiffness was obtained, and the value was strongly influenced by the lower stiffness (of the spring). Analysis of the equivalent stiffness value on each side of the spool showed that it was strongly influenced by the smaller value, in this case, the stiffness of the spring. This stemmed from the stiffness elements being connected in a series. The obtained data revealed that the introduction of shims did not significantly change the amplitude of the spool vibration acceleration. The shim material that exhibited proper isolation efficiency was selected due to theoretical analyzes and numerical considerations using the model of a single-mass system with linear or non-linear characteristics. Limitations resulting from the characteristics of the proportional electromagnets and the design solution of the directional valve should be considered. Typical adjustable-stroke proportional electromagnets used in single-stage proportional valves generate a maximum force of approx. 15–20 N [13] and a stroke of approx. 4.5 mm. These values limit the stiffness of the shim, because the electromagnet's control force must be greater than the sum of the other forces, i.e., friction force in the spool pair, inertia force of the spool, and the associated fluid, force of equivalent stiffness of centering springs and shims, and hydrodynamic force. The use of a shim with excessive stiffness may lead to the sum of the forces acting on the spool becoming greater than the electromagnet's force.

5.3   Theoretical Analysis for the Possible Reduction of Valve Vibrations

The experimental tests in privius sub-chapter show that when choosing a method to reduce the effect of external mechanical vibration on a hydraulic directional control valve and its spool, the limits of the dimensions of vibration isolators, as well as their stiffness and damping, must be taken into account. Geometric limitations stem from the installation conditions, i.e., the available space in which the insulator can be installed (e.g., set of elastomer shims). The constraints of the insulator stiffness result from the maximum power generated by the spool control element, e.g., a proportional electromagnet. In this section, a generalized theoretical consideration of spool and valve body vibrations is used. The considerations excluded the interaction of the valve body and the mounting bracket (i.e., interaction with securing catches, with no friction between the valve body and the mounting bracket).

For theoretical considerations, the single-mass dynamic model can be used and performed in two ways. The first method reduces the vibrations of the valve body, which is an excitation for the spool, and the second method consists of the relative vibration reduction of the spool inside the body [14]. In both cases, a single-mass model with one degree of freedom is employed, in which m represents the mass of the directional valve or the mass of the spool, depending on the case considered, and c and k are the equivalent stiffness and equivalent damping of flexible elements (shims) between the directional valve body and the holder or equivalent stiffness of elastic shims and centering springs and equivalent damping of elastic shims of the spool pair, respectively—Fig. 5.23.

A body of mass m is subjected to a kinematic excitation with a fixed frequency and amplitude and given as the following harmonic equation:

$$w = w_0 \sin(\omega t) \tag{5.15}$$

where w_0—vibration amplitude [m], $\omega = 2\pi f t$ [rad/s], f—frequency [Hz], t—time [s].

The absolute motion of the body (distributor body or spool) is described by the following equation:

Fig. 5.23 Model of a vibrating system with one degree of freedom

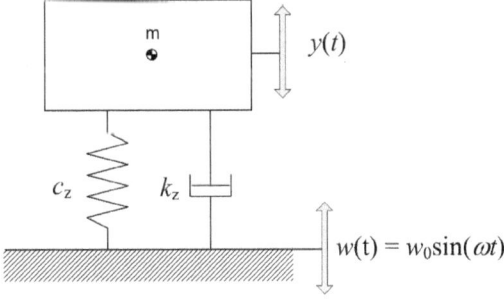

$$m\ddot{x} + k_z(\dot{x} - \dot{w}) + c_z(x - w) = 0. \tag{5.16}$$

However, the displacement of a body with mass m relative to the vibrating base (e.g., directional valve body) can be described as follows:

$$y = x - w. \tag{5.17}$$

Taking into account Eq. (5.16) the relative motion and substitution (5.17), formulated the Eq. of relative motion:

$$m\ddot{y} + m\ddot{w} + k_z\dot{y} + c_zy = 0. \tag{5.18}$$

Due to the fact that w is a known function of time, Eq. (5.18) can be written as:

$$\ddot{y} + 2h_r\dot{y} + w_0^2y = -\ddot{w}(t) \tag{5.19}$$

Due to the excitation being a harmonic function, the following equation can be written:

$$\ddot{y} + 2h_r\dot{y} + w_0^2y = w_0\omega^2\sin(\omega t), \tag{5.20}$$

where:

$$h_r = \frac{k}{2m} \tag{5.21}$$

$$w_0^2 = \frac{c}{m} \tag{5.22}$$

The relatively forced vibrations of the directional valve spool are described by the following equation:

$$y(t) = B_aw_0 \sin(\omega t + \Phi) = B_aw(t + \tau) \tag{5.23}$$

The oscillations of the spool are proportional to the vibration of the body, but are shifted by time $\tau = \Phi/\omega$:

$$y(t) = B_a \cdot w_0 \cdot \sin(\omega t + \Phi) = B_a \cdot w(t + \tau), \tag{5.24}$$

$$w(t) = w_0\sin(\omega t) \tag{5.25}$$

$$w(t + \tau) = w_0 \sin(\omega(t + \tau)), \tag{5.26}$$

$$\sin(\omega t + \Phi) = \sin(\omega(t + \tau)), \tag{5.27}$$

$$\Phi = \omega \cdot \tau, \tag{5.28}$$

where τ—shift in the time plane [s], Φ—phase shift angle [rad].

The coefficient B_a is called the transmission factor, as described by Piersol and Paez [15]:

$$B_a = \frac{\left(\frac{\omega}{\omega_0}\right)^2}{\sqrt{\left(1 - \frac{\omega^2}{\omega_0^2}\right)^2 + 4\frac{h_r^2}{\omega_0^2}\frac{\omega^2}{\omega_0^2}}} \tag{5.29}$$

or

$$B_a = \frac{y_0}{w_0} \tag{5.30}$$

where y_0—amplitude of the relative displacement of the spool [m], w_0—amplitude of the valve body vibration [m].

The value of the tangent of the phase shift angle between the excitation (separator body vibrations) and the response (spool vibration) depends on the ratio between the excitation frequency ω, its natural frequency ω_0 of the system, and the damping value. The steady relative motion described by Eq. (5.18) with subsequent notations, can be expressed as:

$$tg\Phi = \frac{2 \cdot \gamma \cdot \frac{\omega}{\omega_0}}{1 - \left(\frac{\omega}{\omega_0}\right)^2} \tag{5.31}$$

where γ—dimensionless damping coefficient.

Phase shift angle Φ is close to zero at small ratios ω/ω_0 (slow excitation)—case 1. For high ratios ω/ω_0 (fast excitation), the angle is close to π—case 2. For resonance, this angle is equal to $\pi/2$—case 3. This is graphically illustrated in Fig. 5.24. For a steady-state vibrating motion:

Fig. 5.24 Phase shift angle Φ as a function of the ratio between frequency and different values of the dimensionless damping coefficient γ

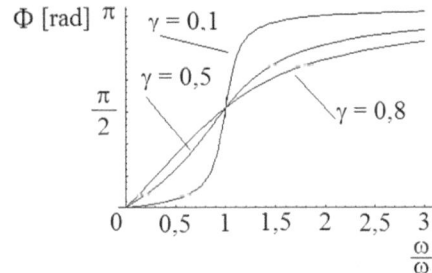

$$\frac{y}{w_0} = \frac{\left(\frac{\omega}{\omega_0}\right)^2}{\sqrt{\left(1 - \frac{\omega^2}{\omega_0^2}\right)^2 + \left(2\frac{h_r}{\omega_0}\frac{\omega}{\omega_0}\right)^2}} \cdot \sin(\omega t - \Phi) \tag{5.32}$$

these three characteristic cases and the form of the steady-state vibration amplitude can be shown as follows:

case 1: slow excitation, i.e. $\frac{\omega}{\omega_0} \ll 1$ and $\Phi \approx 0$:

$$\frac{y}{w_0} \approx 0 \tag{5.33}$$

$$y_0 \approx 0. \tag{5.34}$$

case 2: fast excitation, i.e. $\frac{\omega}{\omega_0} \gg 1$ and $\Phi \approx \pi$:

$$\frac{y}{w_0} \approx 1 \cdot \sin(\omega_0 t - \pi); \tag{5.35}$$

$$y_0 \approx w_0 \tag{5.36}$$

case 3: excitation at the resonant frequency, i.e. $\frac{\omega}{\omega_0} = 1 i \Phi = \frac{\pi}{2}$:

$$\frac{y}{w_0} \approx \frac{1}{2\frac{h_r}{\omega_0}} \cdot \sin\left(\omega_0 t - \frac{\pi}{2}\right); \tag{5.37}$$

$$y_0 \approx \frac{w_0}{2\frac{h_r}{\omega_0}} \tag{5.38}$$

if ratio $\frac{h_r}{\omega_0} \approx 0$, then $y_0 \to \infty$.

where y_0—amplitude of relative steady-state vibrations [m], w_0—excitation amplitude [m].

For a given frequency of kinematic excitation, a 3D graph can be plotted to show the relationship between the transmission coefficient B_a and the equivalent stiffness (centering springs and flexible shims) and equivalent damping inside the directional valve body. Figure 5.25 presents such a relationship for the assumed parameters: spool mass $m = 0.18$ kg, excitation frequency $f = 80$ Hz.

In order to reduce the relative vibration of the spool, the equivalent stiffness or equivalent damping inside the valve body should be selected so that $B_a < 1$. Assuming the equivalent damping inside the directional valve body is equal to 25 kg/s and $f = 80$ Hz, the value of the equivalent stiffness can be determined for which $B_a < 1$—Fig. 5.26.

Fig. 5.25 Dependence of the transmission coefficient B_a on the equivalent stiffness and damping inside the directional valve

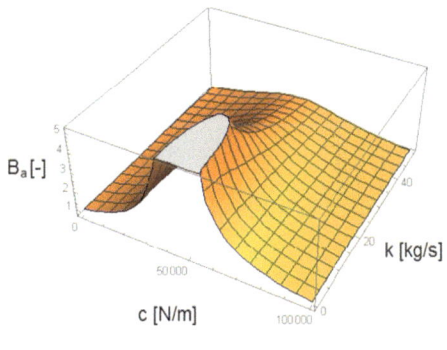

Fig. 5.26 Coefficient B_a as a function of the equivalent stiffness for the given damping and excitation parameters

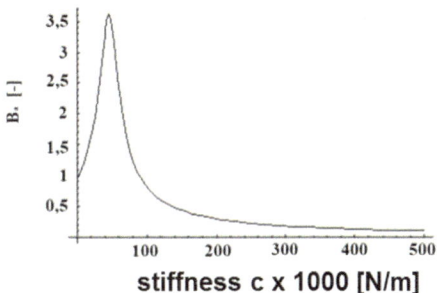

Figure 5.26 shows that the equivalent stiffness inside the directional valve is greater than 100.000 N/m, which avoids spool resonance for these excitation parameters. However, as mentioned earlier, in terms of the control forces in the proportional valve, stiffness is not feasible due to the maximum forces generated by the proportional solenoids (typically approx. 15 N). Assuming that the spool is at full stroke at 4.5 mm, the maximum equivalent stiffness value is 3333 N/m (excluding the share of other forces, which must also be overcome by the control force from the electromagnets).

Hence, assuming that the equivalent stiffness equals 3000 N/m and the excitation parameters are the same, it is possible to analyze the effect of the equivalent damping on coefficient B_a—Fig. 5.27.

Figure 5.27 shows that the equivalent damping must be greater than 30 kg/s for the given equivalent stiffness.

Fig. 5.27 Coefficient B_a value as the function of equivalent damping k for the given stiffness c and the excitation parameters

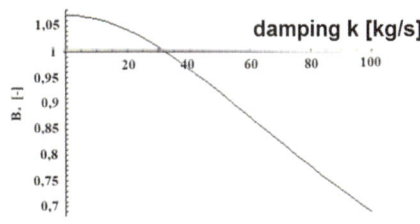

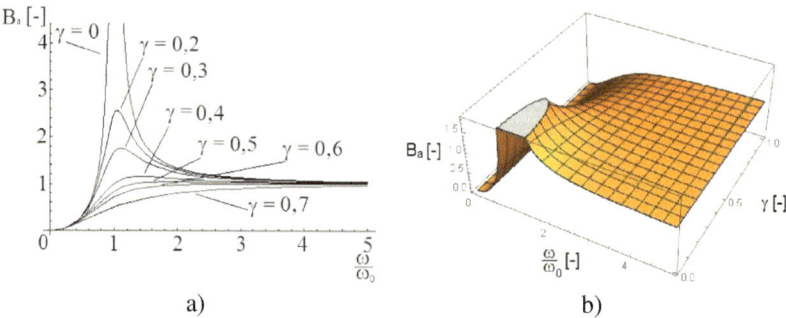

a)   b)

Fig. 5.28 Dependence of coefficient B_a on the ratio of excitation frequency ω and the natural frequency ω_0: **a** for a fixed value of the dimensionless damping coefficient γ.; **b** and dimensionless damping coefficient γ

A graphical interpretation of the dependence of coefficient B_a on the frequency ratio for a fixed dimensionless damping coefficient may be helpful when searching for the expected properties of materials to isolate the vibrations of a spool $\gamma = \frac{k}{2m\omega_0}$—Fig. 5.28a or, in the case of a variable dimensionless damping coefficient, Fig. 5.28b.

The directional valve body can be analyzed in a similar way to minimize the absolute vibration amplitude of the directional valve body [16]:

$$x_0 = w_0 \cdot \sqrt{\frac{1 + \left(2\gamma \frac{\omega}{\omega_0}\right)^2}{\left(1 - \left(\frac{\omega}{\omega_0}\right)^2\right)^2 + \left(2\gamma \frac{\omega}{\omega_0}\right)^2}} \rightarrow \min \qquad (5.39)$$

where w_0—excitation amplitude [m], ω—excitation frequency [rad/s], ω_0—natural frequency of the system [rad/s], γ—dimensionless damping coefficient.

It should be noted that, as with relative vibrations, for steady-state motion, the phase shift angle Φ varies from 0 to π. Its fastest changes can be observed when the excitation frequency approx. equals the natural frequency of the vibrating mass—Fig. 5.29. In this case, this angle is expressed by the following Eq. [16]:

$$\Phi = \text{arc tg} \frac{2\gamma \frac{\omega}{\omega_0}}{\frac{1}{\left(\frac{\omega}{\omega_0}\right)^2} - 1 + 4\gamma^2} \qquad (5.40)$$

For steady-state absolute motion, it is possible to define a coefficient related to the ratio between the amplitude of the absolute displacement of the directional valve body x_0 and the amplitude of substrate vibration w_0. This coefficient is called the transmission factor or the amplification factor T_a [15, 16] to distinguish it from factor B_a for relative motion:

Fig. 5.29 Phase shift angle δ as a function of the ratio between frequency and different values of the dimensionless damping coefficient γ

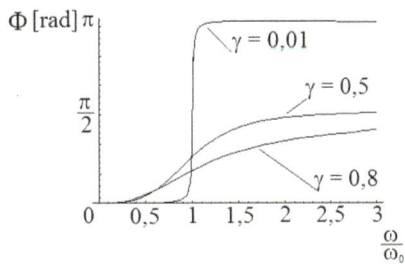

$$T_a = \frac{x_0}{w_0} = \sqrt{\frac{1 + \left(2\gamma\frac{\omega}{\omega_0}\right)^2}{\left(1 - \left(\frac{\omega}{\omega_0}\right)^2\right)^2 + \left(2\gamma\frac{\omega}{\omega_0}\right)^2}} \tag{5.41}$$

The expected properties of isolating materials for distributor body vibrations can be interpreted graphically using the relationship between coefficient T_a and the frequency ratio of a fixed dimensionless damping γ (Fig. 5.30a) or in the case of varying the dimensionless damping coefficient, Fig. 5.30b.

In addition, the effectiveness of vibration isolation can be defined by the following relationship [17]:

$$\varepsilon = \left(1 - \frac{x_0}{w_0}\right)100\% \tag{5.42}$$

In this case, parameters c and k are selected to efficiently determine the vibration isolation from Eq. (5.41), which can be as high as possible or close to 100% (Fig. 5.31).

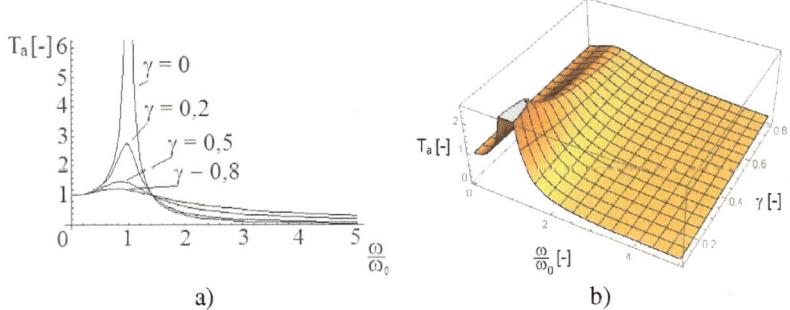

a)   b)

Fig. 5.30 Dependence of coefficient T_a on the ratio of excitation frequency ω and the natural frequency ω_0: **a** for a fixed value of the dimensionless damping coefficient γ; **b** and dimensionless damping coefficient γ

Fig. 5.31 Dependence of coefficient ε on the ratio of excitation frequency ω and the natural frequency ω₀ and dimensionless damping coefficient γ

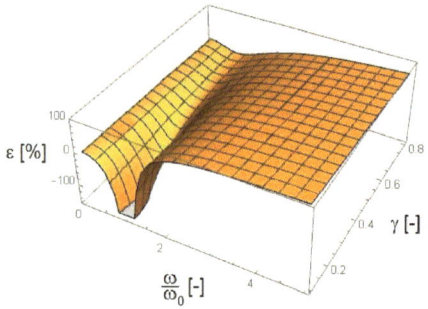

The theoretical considerations presented can be used to select and evaluate the parameters of a passive vibration isolation system and allow the limitations of such a system to be determined.

5.4   Examples of Using Approximate Analytical Methods for the Analysis of Non-Linear Valve Vibration Reduction Systems

In engineering practice, the analysis of vibrating systems is often based on considerations of linear systems. However, more detailed and accurate results can be obtained by assessing the non-linearities that sometimes occur. In mechanical systems, sources of non-linearities can be elastic or damping elements. In some cases, the weight of the system is variable. Damping can be proportional not only to the first power of velocity but also to the square of velocity v^2. Additionally, the spring stiffness constant can change with the deflection of the spring. Herein, only selected literature is mentioned [12, 15, 18, 19], which provides a broad analysis of the vibration of non-linear systems. In this section, in relation to the literature (where a specific topic is elaborated), selected models with non-linear damping or non-linear spring are presented. Selected models and approximate solving methods may prove useful in the search and selection of appropriate materials to reduce the impact of external mechanical vibration on the valve body or its control element.

An important feature of the system related to the non-linearity of the spring is the natural frequency varying as a function of the amplitude. This is different than in systems with non-linear damping, where the natural frequency changes insignificantly due to damping changes. In systems with non-linear springs, even a small change in the amplitude can cause significant changes in the natural frequency of the system. Those with non-linear springs, harmonic, and sub-harmonic deflection components also appear. Analysis of a vibrating system with non-linear springs has shown that it is possible to obtain harmonic components of motion with values higher or lower than the frequency of the excitation force.

It is convenient to analyze vibrating systems with non-linearities using the Schwesinger method Wereley [19]. This is an approximate method based on a one-term approximation, where non-linear spring or non-linear damping can be considered for the vibrations of systems with variable mass. Additionally, it is assumed that the vibrations are harmonic, and the deflections can be described by the following equation:

$$x(t) = x_0 \sin(\omega t).$$
(5.43)

This assumption causes a certain inaccuracy of the method because only one component is considered while excluding the harmonic components of higher orders. This method begins with the formulation of the differential Eq. of forces acting on a body of mass m. Schwesinger introduces the concept of a substitute force causing disturbances in the operation of a body of mass m. For harmonic motion, this force would be zero for a linear system. However, harmonics of higher orders are present in the response of a non-linear system and in this case, the forced deflection can be expressed as:

$$x(t) = x_0 \sin(\omega t) + x_{02} \sin(2\omega t) + x_{03} \sin(3\omega t) + \ldots + x_{0n} \sin(n\omega t)$$
(5.44)

The method assumes a single-body approximation, hence, the forces occurring due to harmonics of higher orders are neglected (5.44). The equivalent force acting on the body concentrates the harmonic forces of higher orders. Reports have shown that the method defines the integral of the square of the introduced equivalent force and seeks its minimum. Assuming an amplitude value of x_0, it is possible to determine the force for which the integral reaches its minimum. The force components F_1 and F_2 are obtained as follows:

$$F_1 = \frac{1}{\pi} \int_0^{2\pi} \left[F(x) - m \cdot \omega^2 \cdot x_0 \sin(\omega t) + \frac{dm}{dx} \cdot \omega^2 x_0^2 \cos^2(\omega t) \right] \cdot \sin(\omega t) d(\omega t)$$
(5.45)

$$F_2 = \frac{1}{\pi} \int_0^{2\pi} F(\dot{x}) \cdot \cos(\omega t) d(\omega t).$$
(5.46)

The overall force is expressed by the following equation:

$$F^2 = F_1^2 + F_2^2,$$
(5.47)

and the phase shift is described by the following equation:

$$\Phi = \text{arc tg} \frac{F_2}{F_1}$$
(5.48)

The characteristic equation is obtained after transformations based on Eq. (5.47). Using this Eq., and adopting the values of the vibration amplitude, the corresponding frequencies are obtained. In this way, a curve is obtained that describes the steady-state vibration of a body with mass m and springs of non-linear characteristics.

The method can evaluate how effectively a non-linear system isolates the vibration of the directional valve, which is affected by harmonic external excitation. It is possible to assess the impact of the non-linear system on the reduction of the valve body vibrations excited by the harmonic force. For example, the stiffness characteristic of a vibration isolator can be expressed by an Eq. in which the stiffness force depends on the deflection of the vibration isolator in the first and third power with accuracy to constants characterizing the stiffness:

$$F(x) = c_1 \cdot x + c_3 \cdot x^3 \tag{5.49}$$

where c_1 and c_3 are spring constants with units [N/m] and [N/m^3], respectively, and the damping of the vibration isolator is proportional to the velocity in the first power:

$$F(\dot{x}) = k \cdot \dot{x} \tag{5.50}$$

where k is the damping constant [kg/s], and the relationships for the force components F_1 (5.51) and F_2 (5.52) can be written as follows:

$$F_1 = \frac{1}{\pi} \cdot \int_0^{2\pi} \left(c_1 \cdot x_0 \cdot \sin(\omega t) + c_3 \cdot x_0^3 \cdot \sin^3(\omega t) - m \cdot x_0 \cdot \omega^2 \cdot \sin(\omega t) \right) \cdot \sin(\omega t) d(\omega t) \tag{5.51}$$

$$F_2 = \frac{1}{\pi} \int_0^{2\pi} k \cdot x_0 \cdot \omega \cdot \cos^2(\omega t) d(\omega t) \tag{5.52}$$

Continuing the procedure by the Schwesinger method, and adopting the system parameters such as the mass of the vibrating directional valve m = 4.5 kg, the vibration isolator stiffness constants: c_1 = 86,000 N/m, c_3 = 86,00000 N/m^3, the amplitude of the harmonic excitation force F_0 = 90 N and the damping constant k_1 = 80 kg/s, one can determine the characteristic Eq. of the steady-state vibration of the directional valve. After assuming the directional valve vibration amplitude, the frequency values can be determined—Table 5.5.

A comparison of linear and non-linear vibration isolating systems is shown in Fig. 5.32. For the linear vibration-isolating system, the value of the stiffness constant c = 86,000 N/m and the damping constant k = 160 kg/s is assumed. In addition, from Eq. (5.53), the amplitude of vibration fixed for the linear system [16] is determined:

$$x_0 = \frac{F_0}{c} \cdot \frac{1}{\sqrt{\left(1 - \left(\frac{\omega}{\omega_0}\right)^2\right)^2 + \left(\frac{k \cdot \omega}{m \cdot \omega_0^2}\right)^2}} \tag{5.53}$$

Table 5.5 The values of the steady-state vibration amplitude of a non-linear system and the frequency values determined by the Schwesinger method

x_0 [m]	F_1 [Hz]	F_2 [Hz]
0.0081	21.93	22.00
0.0078	21.54	22.38
0.0070	21.01	22.78
0.0060	20.60	23.20
0.0050	20.25	23.68
0.0040	19.24	24.31
0.0030	17.98	25.25
0.0020	15.32	26.94
0.0015	12.17	28.50
0.0011	4.88	30.60

As shown in Fig. 5.32, the insertion of a vibration isolator with a non-linear stiffness characteristic significantly increases the steady-state vibration amplitude for the resonant frequency of the vibrating directional control valve.

To analyze the effects of damping non-linearity with linear stiffness characteristics on the effectiveness of vibration isolation, the concept of viscous equivalent damping can be used. Equivalent (substitute) viscous damping is introduced by replacing other forms of energy dissipation, such as dry friction and friction proportional to the square of velocity.

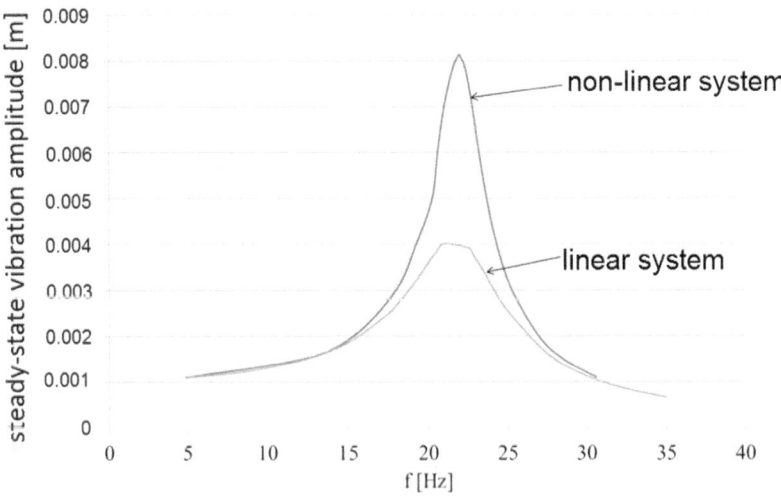

Fig. 5.32 Steady-state vibration amplitude of a system with non-linear stiffness characteristics and linear damping, and vibration amplitude of a system with linear stiffness characteristics and linear damping

The system is affected by equivalent viscous damping to the same extent as by all the forms of equivalent damping [15, 19]. In this method, the non-linear differential Eq. is replaced by a linear Eq., thus, the description of the system becomes linearized. The method assumes that a harmonic force acts on the body.

For example, the damping characteristics can be expressed as follows:

$$F(\dot{y}) = k \cdot \dot{y} = -y_0 \cdot k \cdot \omega \cdot \sin(\omega t), \tag{5.54}$$

where $\dot{y}$ is the time derivative of the relative deflection $y = y_0 \cdot \cos(\omega t)$, then it is possible to determine the work of the force per period:

$$W_T = k \cdot y_0^2 \cdot \omega \cdot \int_0^{2\pi} \sin^2(\omega t) d(\omega t) = k \cdot y_0^2 \cdot \omega \cdot \pi, \tag{5.55}$$

where W_T—work per a single period.

The general formula for the work of a force per period can be written as follows [19]:

$$W_T = \int_0^{2\pi} F(\dot{y}) \cdot \dot{y} dt \tag{5.56}$$

From the comparison of relationships (5.56) and (5.57), the equivalent damping coefficient for a damper with non-viscous drag is obtained:

$$k_{zast} = \frac{1}{y_0^2 \cdot \omega \cdot \pi} \cdot \int_0^{2\pi} F(\dot{y}) \dot{y} dt \tag{5.57}$$

However, the method has some limitations and is, therefore, an approximate method. Importantly, the motion is assumed to be harmonic. The applicability of this method is limited to cases where damping does not distort the sine wave of the moving body.

The method allows a more general conclusion stating that for many systems with non-linear damping, the steady-state vibration curve of the body is smoother. Such an observation can be made for a vibrating single-mass system with a vibration isolator with a linear stiffness characteristic and damping proportional to v_2. In this case, a body of mass m is affected by external vibration, and the body is placed on a vibration isolator with linear stiffness characteristics and damping expressed as follows:

$$F(\dot{y}) = \lambda \cdot \dot{y}^2 \tag{5.58}$$

where λ is constant, and $\dot{y}$ is the relative velocity.

By writing the relative displacement of a body with mass m as follows:

$$y = y_0 \cdot \cos(\omega t) \tag{5.59}$$

the work of a force per period is determined for the damper whose equivalent resistance is described by Eq. (5.60):

$$W_T = 2,66 \cdot \omega^2 \cdot y_0^3 \cdot \lambda \tag{5.60}$$

The equivalent damping coefficient is obtained by comparing the relationships (5.61) and (5.62):

$$k_{zast} = 0,8488 \cdot y_0 \cdot \omega \cdot \lambda \tag{5.61}$$

To obtain the form of the dimensionless damping coefficient γ, the form of the equivalent damping coefficient should be substituted into the following relationship:

$$\gamma = \frac{k_{zast}}{2m \cdot \omega_0} = 0,4244 \cdot y_0 \cdot B \cdot \frac{\omega}{\omega_0} \tag{5.62}$$

where $B = \frac{\lambda}{m}$.

Then, the determined formula for the dimensionless damping coefficient can be introduced into Eq. for the relative motion of a system with one degree of freedom with a linear stiffness characteristic and viscous damping:

$$\frac{y_0}{w_0} = \frac{\left(\frac{\omega}{\omega_0}\right)^2}{\sqrt{\left(1 - \frac{\omega^2}{\omega_0^2}\right)^2 + \left(2\gamma\frac{\omega}{\omega_0}\right)^2}} \tag{5.63}$$

to obtain the expression:

$$\frac{y_0}{w_0} = \frac{\left(\frac{\omega}{\omega_0}\right)^2}{\sqrt{\left(1 - \frac{\omega^2}{\omega_0^2}\right)^2 + 0,72 \cdot \left(y_0 \cdot B \cdot \frac{\omega^2}{\omega_0^2}\right)^2}} \tag{5.64}$$

using substitution $z = y_0^2$ and transforming it, geted:

$$0.72 \cdot B^2 \cdot \left(\frac{\omega}{\omega_0}\right)^4 \cdot z^2 + \left(1 - \left(\frac{\omega}{\omega_0}\right)^2\right)^2 \cdot z - \left(\frac{\omega}{\omega_0}\right)^4 \cdot w_0^2 = 0 \tag{5.65}$$

After parameterization of Eq. (5.65), it can be solved with respect to y_0. A formula is obtained for the dependence of the amplitude of relative vibration determined on the frequency ratio ω/ω_0. If positive discriminant is assumed, then the real roots of Eq. (5.66) are such numbers as z_1 and z_2 that satisfy the relationship:

$$z_{1,2} = -\frac{0.694}{B^2} - \frac{0.694}{\left(\frac{\omega}{\omega_0}\right)^4 \cdot B^2} + \frac{1.389}{\left(\frac{\omega}{\omega_0}\right)^2 \cdot B^2}$$

$$\pm \frac{0.694 \cdot \sqrt{\left(1 - \left(\frac{\omega}{\omega_0}\right)^2\right)^4 + 2.88 \cdot \left(\frac{\omega}{\omega_0}\right)^8 \cdot B^2 \cdot w_0^2}}{\left(\frac{\omega}{\omega_0}\right)^4 \cdot B^4} \tag{5.66}$$

Equations (5.65) and (5.66) show that the relative motion of a body of mass m is not linear with respect to the inclination w_0, as in the case of the linear system described by Eq. (5.32).

As determined from the experimental tests, despite a significant reduction of up to 70% in the vibration acceleration in the isolation range (Fig. 5.13), the isolation in the form of springs causes the vibration in the resonance area to increase by nearly 70%. Reports have shown that this will result in the pulsation components having frequencies of 10–20 Hz in the amplitude-frequency spectrum and may generate mechanical vibration with corresponding frequencies and infrasound. Another feasible method of reducing the vibration of the directional valve body and, consequently, of the spool, is to place the directional valve on elastic shims. The experimental data shown in Figs. 5.19 and 5.22 indicate that despite the materials reducing the vibration of the spool or the body, they are expected to be more effective at vibration isolation.

Our considerations also indicate that the vibration of the directional valve spool can be reduced by inserting a shim (of a material with high stiffness and damping) inside the directional valve body, between the body and the centering springs. In the case of single-stage electrically controlled directional valves (e.g., proportionally), this approach is limited due to the maximum values of the control forces created by the proportional solenoids and the required stroke of the spool.

The presented linear and non-linear mathematical models of vibration isolation can be used to select the appropriate materials in terms of isolation effectiveness. For the isolation of vibration of the directional valve body, materials can be selected based on Eqs. (5.39), (5.41), (5.42), (5.53), and (5.63), and for the isolation of the spool's vibration Eqs. (5.2), (5.3), (5.29), and (5.32) may be useful.

5.5   Reduce Vibration in Hydraulic Systems by Biomimetic Approach Applied for Pipelines

Examining contemporary energy-efficient systems and technologies within hydraulic drive mechanisms, a study [20] underscores that a key strategy to diminish power consumption and enhance hydraulic drive efficiency involves the reduction of dynamic loads, specifically pressure pulsations. As outlined by Stosiak et al. [21], typical approaches to mitigate dynamic loads encompass the incorporation of active damping elements into the hydraulic drive. Amirante et al. [22] further details that hydraulic accumulators and various fluid flow dampers are widely utilized for this purpose. Nevertheless, Kitajima et al. [23] highlights instances where integrating damping elements into the hydraulic drive may not be practical or logical. For instance, hydraulic accumulators and pulsation dampers could pose challenges in mobile machine hydraulic systems due to their size or inability to operate within the required frequency range, and in some cases, might even yield adverse effects, as indicated by Kitajima et al. [23]. Proposals such as pumps with built-in pulsation dampers, as suggested by Chai et al. [24], prove suitable for mobile systems but might encounter integration difficulties owing to the substantial space requirements of the dampers. Additionally, Shen et al. [25] points out that active vibration compensation systems have a drawback—they introduce energy into the hydraulic system through the actuator, potentially compromising system stability. Consequently, Shen et al. [25] proposes the exploration of alternative passive damping components to attenuate pressure pulsations within the hydraulic drive.

As per references Bach et al. [26, 27], exploring biomimetic approaches holds promise for uncovering innovative strategies to mitigate pressure pulsations in hydraulic drives. The authors propose leveraging the pulsatile nature of flows delivered by pumps in nature and drawing inspiration from the nonlinear viscoelasticity observed in the circulatory systems of vertebrates. Bach et al. [26] specifically draws parallels between the damping principle inherent in vertebrates' circulatory systems and existing technical solutions, suggesting that the nonlinear viscoelastic properties of arterial vessel walls present significant potential for enhancing current technical dampers or devising novel biomimetic damping solutions, as highlighted in Camasão et al. [28].

Furthermore, Shadwick [29] elucidates that the passive expansion and elastic recoil of arterial walls, a characteristic of major vessels, contribute to the reduction of pressure pulsation magnitudes in the circulatory system and promote a more uniform blood flow. These vessel walls, composed of elastin, collagen fibers, and smooth muscle cells, constitute a composite material, as emphasized by Wolinsky and Glagov [30]. According to Shadwick [29], a portion of the pulse energy is dissipated through the viscoelastic component of these vessel walls, resulting in approximately 15–20% strain energy loss during each cycle. This damping mechanism shares similarities with expansion hoses and inline hydraulic accumulators.

However, conducting a precise dynamic analysis of hydraulic driving processes poses challenges owing to the nonlinear characteristics arising from the compressibility of oil and the nonlinearity inherent in both the pump and hydraulic high-pressure hoses. Consequently, it becomes imperative to undertake experimental investigations aimed at optimizing the efficiency of hydraulic drives, particularly by exploring diverse types of high-pressure hoses. This paper emphasizes the significance of examining the potential to restrict pressure pulsation amplitudes and minimize energy loss in flexible hoses, contingent upon the specific structure of the hose wall.

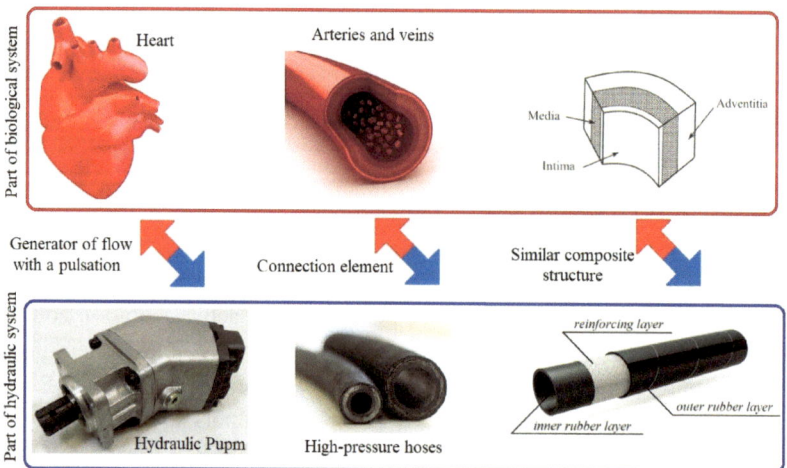

Fig. 5.33 Hydraulic drive vibration damping based on biomimetic approach

Biomimetics, a process involving the emulation of biological systems for the creation of human-made technologies, is finding applications in hydraulic drive science to foster the development of environmentally friendly and sustainable designs, materials, and techniques, as illustrated in Fig. 5.33, based on Karpenko et al. [31]. The figure illustrates the correlation between biological and hydraulic components. Designers draw inspiration from nature's most efficient solutions when formulating their concepts. In hydraulic systems, analogous to the positive displacement pump generating pulsatile flow, living organisms' blood circulation systems rely on the heart as the source of pulsatile flow.

In typical hydrostatic systems, the pulsation spectrum's fundamental component falls within the 150–350 Hz range. For living organisms, the frequency of blood pressure pulsations depends on the organism's size, following a general rule that smaller organisms exhibit higher frequencies (e.g., approximately 1 Hz for humans, 0.4 Hz for elephants, and 12 Hz for mice). This pulsating flow of the working fluid (hydraulic fluid or blood) and the impedance of the hydraulic or blood system contribute to pressure pulsations and, consequently, to the time-varying, harmonic loading of hose or vein walls.

The analogy extends to elements transporting the working fluid, with pipes (rigid or flexible) in hydraulic systems mirroring the veins, aorta, and arteries in circulatory systems. Just as the veins in living organisms comprise three layers (outer membrane, endothelium, muscular layer), designers of high-pressure hydraulic hoses explore innovative materials and conductor designs. Contemporary hydraulic lines, often composed of composite materials and featuring three layers, can transmit high pressures while concurrently mitigating fluid pressure pulsations compared to rigid hoses. In hydraulic systems, where fluid is conveyed to hydraulic receivers such as motors and cylinders, the biological analogy aligns with muscles.

The overarching concept aims to enhance hydraulic drive efficiency through experimental research and a biomimetic approach that leverages various high-pressure hoses. This approach also incorporates a biological analogy to dampen pressure pulsations within hydraulic drives. The idea of harnessing the nonlinear viscoelastic properties of high-pressure hose walls to enhance dampers in hydraulic drives is proposed with substantial potential, drawing inspiration from biomimetic arterial and venous damping solutions. The passive expansion and elastic recoil of high-pressure hose walls are envisaged to curtail pressure pulsation magnitudes and promote a smoother fluid flow throughout the hydraulic system. Analogous to vessel walls, high-pressure hose walls consist of multiple layers, incorporating materials like rubber, fabric, or steel braid. The underlying concept posits that the viscoelastic component of high-pressure hose walls could dissipate a portion of the fluid pulse energy. The adoption of flexible hoses, as opposed to rigid ones, in hydraulic systems is suggested to enhance the system's capacitance (c_k) by decreasing the equivalent bulk modulus of elasticity (B_z). This relationship between these parameters is expressed through linearized relationships:

$$B_z = \frac{\Delta p}{\frac{\Delta V}{V_0}} [Pa], \tag{5.67}$$

$$c_k = \frac{V}{B_z} \left[\frac{m^5}{N} \right], \tag{5.68}$$

where, Δp—change in pressure, ΔV—change in volume due to pressure increment, V_0—initial volume of liquid enclosed in the hose.

Given the pronounced nonlinearity arising from the compressibility of oil and the nonlinear characteristics inherent in both the pump and hydraulic pipeline, conducting a precise dynamic analysis of the hydraulic driving process poses challenges. Consequently, it is recommended to undertake experimental investigations as a pragmatic approach to enhance the efficiency of hydraulic drives, leveraging diverse types of high-pressure hoses. This empirical approach is deemed necessary to navigate the complexities introduced by the nonlinear dynamics in the hydraulic system.

The experimental investigation encompassed the measurement and analysis of fluid pressure drop within high-pressure hoses, coupled with a vibration analysis contingent upon fluid flow. Employing a two-sample measurement design, the research adhered to the single-sample statistical method outlined in Karpenko and Nugaras [32] for estimating uncertainty in the iterative processing of data measurements. The experimental setup, depicted in Fig. 5.34, involved a test bench specifically designed for researching fluid pulsations within high-pressure hoses and their associated deformations.

Within the context of this concept, one viable option for high-pressure hoses involves the utilization of a composite material comprising rubber, as illustrated in Fig. 5.35a. This material serves as a crucial structural component in any hydraulic drive, facilitating the connection of all elements within a unified operational system. In the ongoing research,

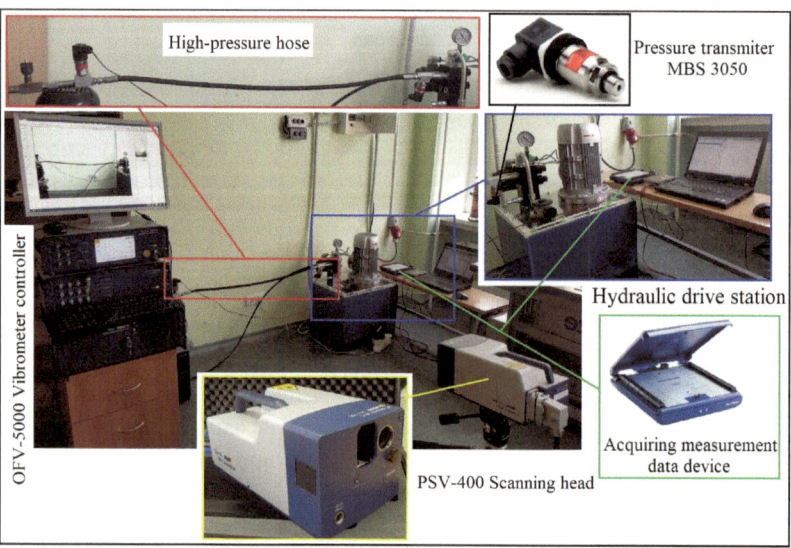

Fig. 5.34 The view of test bench for pipeline measurements

a comparative analysis was conducted to validate the concept, focusing on different types of one-braid high-pressure hoses, namely:

- One braid fabric-reinforced hydraulic hose (TE) standart [33].

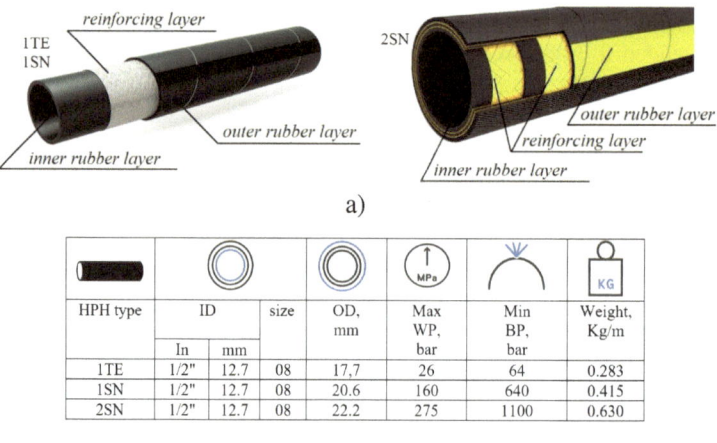

a)

HPH type	ID		size	OD, mm	Max WP, bar	Min BP, bar	Weight, Kg/m
	In	mm					
1TE	1/2"	12.7	08	17,7	26	64	0.283
1SN	1/2"	12.7	08	20.6	160	640	0.415
2SN	1/2"	12.7	08	22.2	275	1100	0.630

b)

Fig. 5.35 Research objects: **a** composition of pipelines (high-pressure hoses); **b** main geometrical parameters of used pipelines (high-pressure hoses)

- Standard one steel braid-reinforced hydraulic hose (SN) standart [34].
- Compact one steel braid-reinforced hydraulic hose (SC) standart [35].

It's worth noting that the SC steel wire-reinforced hydraulic hose is structurally similar to the SN steel wire-reinforced hydraulic hose, with the key distinction being its smaller outside diameters. This characteristic addresses the challenge of compact and narrow installation spaces in certain equipment piping. The research specifically considered high-pressure hoses with a $1/2''$ conditional pass diameter. Figure 5.35a illustrates the material composition of the high-pressure hose layers, and detailed properties of each layer can be referenced in Karpenko [36]. Additionally, Fig. 5.35b presents the main geometrical parameters of the high-pressure hoses employed in the research.

The experimental measurements encompassed multiple readings of fluid pressure at the inlet and outlet of the high-pressure hose, along with several assessments of displacement and velocity of the high-pressure hose surface, including its frequency response. To minimize measurement errors, the presented results represent the average outcomes of multiple readings. Velocity measurements were conducted using a PSV Sensor Head on the surface of the high-pressure hose, as illustrated in Fig. 5.36. Simultaneously, fluid pressure pulsation was gauged at the inlet and outlet of the high-pressure hoses, as depicted in Fig. 5.37.

Additional results obtained from the experimental segment include the frequency response, based on spectrum analyses using the Doppler Effect for the investigated objects. The frequency analysis (Fig. 5.38) indicates that the primary resonance frequencies of pipelines fall within the low-frequency range (0–200 Hz) and the commencement of the middle-frequency range (200–500 Hz). Notably, the primary focus of the research is within the frequency range up to 500 Hz, as this is where the primary resonant modes are observed.

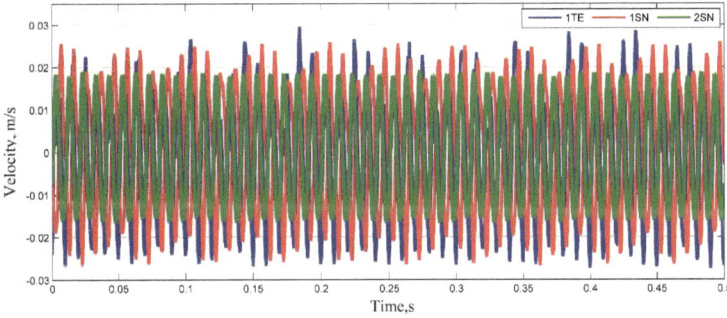

Fig. 5.36 Velocity measurements were conducted using a PSV sensor head on the surface of the pipelines (high-pressure hose)

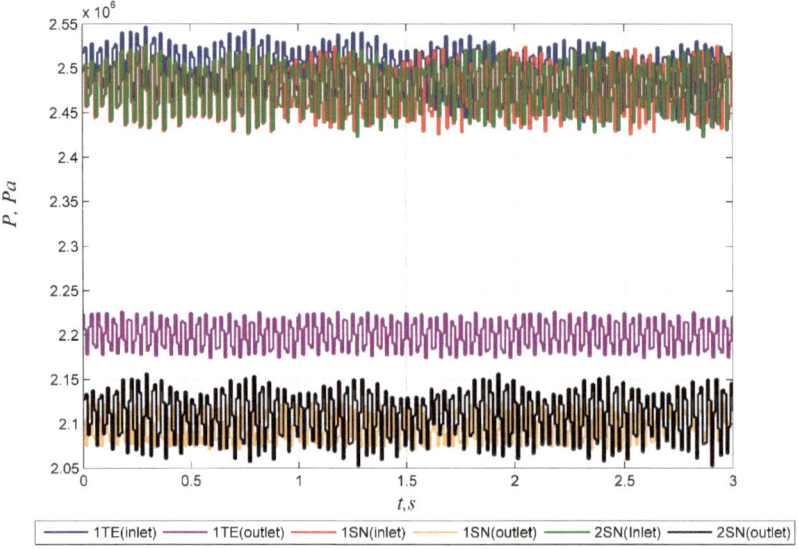

Fig. 5.37 Fluid pressure pulsation gauged at the inlet and outlet of the of the pipelines (high-pressure hose)

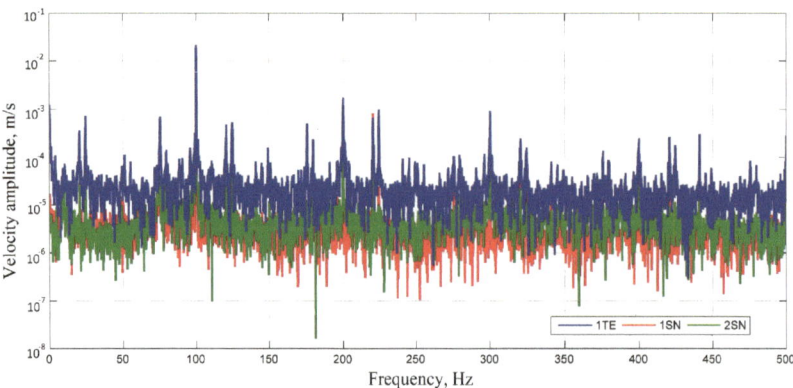

Fig. 5.38 The frequency analysis of the pipelines (high-pressure hose) mechanical vibration

From the frequency analysis, it is apparent that the main and initial resonance frequency of the investigated pipelines is 20 Hz, with harmonics occurring at each 20 Hz interval (20/40…80 Hz, etc.) and a maximum resonance amplitude at 100 Hz. A frequency of 27.93 Hz is also evident in the frequency response of the pipelines outer surface, transmitted from fluid pulsation. It is worth noting the specific frequencies in the low-frequency range (37.5 and 67.5 Hz) for the 1TE pipelines, followed by frequencies of 157.5 and 187.5 Hz, progressing with a step of 30 Hz and continuing into the middle-frequency range. For the 1SN/2SN pipelines, a second set of frequencies begins at 124.04 Hz, with harmonic steps. The existence of these second resonant frequencies is explained by the composite nature of the investigated objects. Based on the obtained spectrum analyses, it appears that 1TE pipelines exhibits superior damping properties across different frequencies compared to 1SN/2SN pipelines.

By frequency's response of fluid flow inside the pipelines (Fig. 5.39) it is seen that the main fluid pressure amplitudes on frequency 27.93, 44.13, 60.33, 72.06 and 127.93, 144.13, 160.33, 172.06 Hz.

The difference in the amplitude of fluid pulsation on the inlet and outlet of pipelines confirmed that pipelines can be used to damp fluid pulsation and reduce fluid pressure losses inside hydraulic lines. Since, stiffness and deformation of pipelines are different, the damping of fluid pulsation between inlet and outlet amplitude constitute ~11.6% for 2SN, ~49.5% for 1SN and ~54.4% for 1TE pipelines.

The measurements suggest that the standard version (SN) of the high-pressure hose exhibits superior performance in mitigating fluid pressure pulsation and dynamic loads within hydraulic lines. This superiority is evidenced by a smaller fluid pressure drop and lower fluid pressure amplitude at the outlet compared to the compact version (SC). Conversely, the best energy-saving characteristics appear to be achievable by employing the version with a textile cord (TE). This observation is likely attributed to the thicker rubber layers in the radial direction, which may contribute to better damping of fluid pulsation. However, it's important to note that this thicker rubber layer may also result in lower stiffness and increased flexibility of the braid layer.

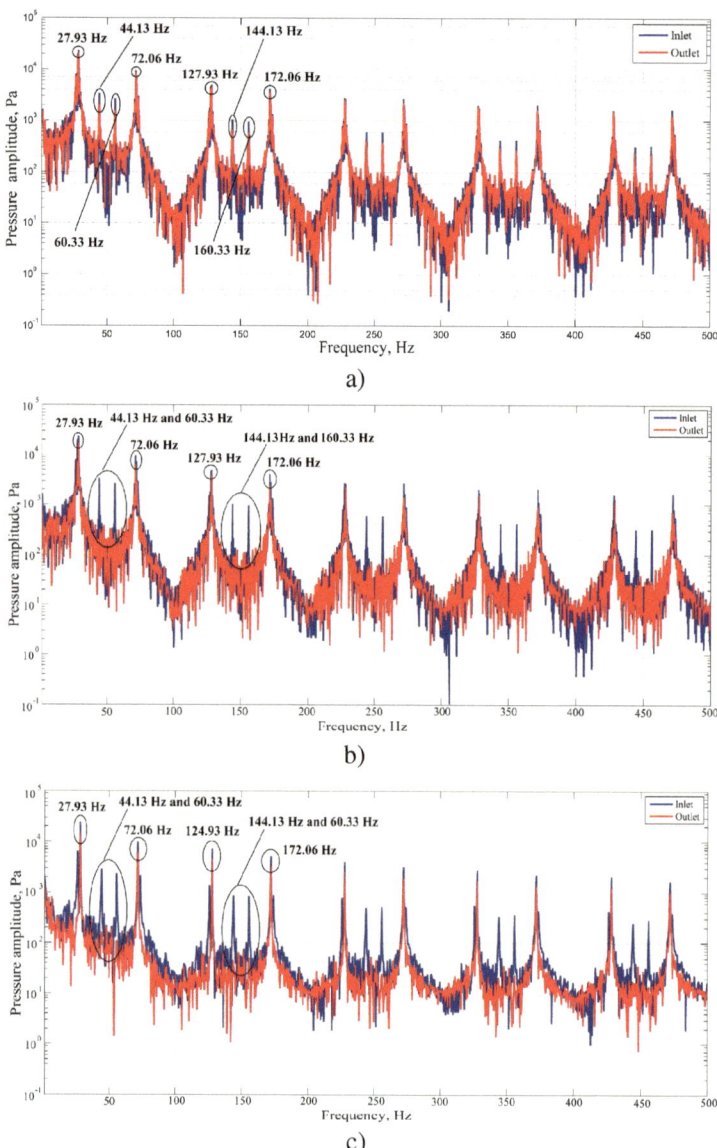

Fig. 5.39 The frequency analysis of fluid flow pulsation inside the of the pipelines (high-pressure hose): **a** 2SN type; **b** 1SN type; **c** 1TE type

References

1. Parker-Hannifin. Catalog MSG-14-2550/US. Technical information of proportional directional control valves series D1FP. 1–6 p. Retrieved from https://www.parker.com/content/dam/Parker-com/Literature/Hydraulic-Valve-Division/hydraulicvalve/Catalog-sections-for-websphere/Proportional-Directional-Control/Catalog--Static-Files/D1FP.pdf
2. EN 60068-2-6 Environmental testing—Part 2–6: Tests—Test fc: Vibration (sinusoidal).
3. PN-EN 60068-2-57:2013-12. (2013). Badania środowiskowe—Część 2–57: Próby—Próba Ff: Wibracje—Metoda zarejestrowanego przebiegu drgań i dudnień sinusoidalnych (50 p.) [In Polish].
4. PN-EN 60068–2–6 Badania środowiskowe – część 2–6: próby – próba fc: wibracje (sinusoidalne)
5. PN-EN ISO 4413. (2011). Napędy i sterowania hydrauliczne – Ogólne wytyczne dotyczące układów (50 p.) [In Polish].
6. ISO 2954:2012. Mechanical vibration of rotating and reciprocating machinery—Requirements for instruments for measuring vibration severity.
7. Directive 2006/42/EC of the European Parliament and of the Council of 17 May 2006 on machinery, and amending Directive 95/16/EC (recast) (182 p.). Retrieved from https://eur-lex.europa.eu/legal-content/EN/TXT/?uri=celex%3A32006L0042
8. Directive 2005/88/EC of the European Parliament and of the Council of 14 December 2005 amending Directive 2000/14/EC on the approximation of the laws of the Member States relating to the noise emission in the environment by equipment for use outdoors (Text with EEA relevance) (173 p.). Retrieved from https://eur-lex.europa.eu/legal-content/EN/TXT/?uri=CELEX%3A32005L0088
9. Directive 2007/30/EC of the European Parliament and of the Council of 20 June 2007 amending Council Directive 89/391/EEC, its individual Directives and Council Directives 83/477/EEC, 91/383/EEC, 92/29/EEC and 94/33/EC with a view to simplifying and rationalising the reports on practical implementation (Text with EEA relevance) (42 p.). Retrieved from https://eur-lex.europa.eu/legal-content/en/ALL/?uri=CELEX%3A32007L0030
10. ISO 1219-1:2016. Fluid power systems and components—Graphical symbols and circuit diagrams—Part 1: Graphical symbols for conventional use and data–processing applications.
11. Walha, L., Jarraya, A., Djemal, F., et al. (2021). *Design and modeling of mechanical systems—V. Lecture notes in mechanical engineering* (912 p.). Springer.
12. Singh, R., & Goyal, VI. (2021). *Modeling & simulation of dynamical systems* (68 p.). LAP Lambert Academic Publishing.
13. Manring, N., & Fales, R. (2019). *Hydraulic control systems* (2nd ed., 376 p.). Wiley.
14. Stosiak, M. (2014). Ways of reducing the impact of mechanical vibrations on hydraulic valves. *Archives of Civil and Mechanical Engineering, 15*(2), 1–9.
15. Piersol, A., & Paez, T. (2009). *Harris' shock and vibration handbook* (6th ed., 1168 p.). McGraw-Hill Handbooks.
16. Wen, B., Huang, X., Li, Y., & Zhang, Y. (2023). *Vibration utilization engineering* (351 p.). Springer.
17. Schmitz, T., & Smith, K. (2021). *Mechanical vibrations. Modeling and measurement* (430p.). Springer.
18. Hutchings, I., & Shipway, P. (2017). *Tribology: Friction and wear of engineering materials* (2nd ed., p. 412). Elsevier.
19. Wereley, N. (2021). *Introduction to vibration in engineering* (204 p.). Cognella Academic Publishing.

20. Karpenko, M., & Bogdevičius, M. (2017). Review of energy-saving technologies in modern hydraulic drives. *Science—Future of Lithuania, 9*(5), 553–558.
21. Stosiak, M., Karpenko, M., Deptuła, A., Urbanowicz, K., Skačkauskas, P., Deptuła, A. M., Danilevičius, A., Šukevičius, Š, & Łapka, M. (2023). Research of vibration effects on a hydraulic valve in the pressure pulsation spectrum analysis. *Journal of Marine Science and Engineering, 11*(2), 1–15.
22. Amirante, R., Distaso, E., & Tamburrano, P. (2014). Experimental and numerical analysis of cavitation in hydraulic proportional directional valves. *Energy Conversion and Management, 87*, 208–219.
23. Kitajima, D., Machimura, H., Munakata, A., Nemoto, M., & Yamauchi, H. (2013). Fluid pressure pulsation damper mechanism and high-pressure fuel pump equipped with fluid pres-sure pulsation damper mechanism. US patent US,83,664,21,B2.
24. Chai, L., Jiao, Z., Xu, Y., & Zheng, H. (2016). A compact design of pulsation attenuator for hydrau-lic pumps. In *Proceedings of IEEE International Conference on Aircraft Utility Systems (AUS)* (pp. 1111–1116).
25. Shen, W., Jiang, J., Su, X., & Karimi, H. (2015). Control strategy analysis of the hydraulic hybrid excavator. *Journal of the Franklin Institute, 352*(2), 541–561.
26. Bach, D., Schmich, F., Masselter, T., & Speck, T. (2015). A review of selected pumping systems in nature and engineering—Potential biomimetic concepts for improving displacement pumps and pulsation damping. *Bioinspiration & Biomimetics, 10*(4), 1–28.
27. Bach, D., Masselter, T., & Speck, T. (2017). Damping of pressure pulsations in mobile hydraulic applications by the use of closed cell cellular rubbers integrated into a vane pump. *Journal of Bionic Engineering, 14*(4), 791–803.
28. Camasão, D., & Mantovani, D. (2021). The mechanical characterization of blood vessels and their substitutes in the continuous quest for physiological-relevant performances, A critical review. *Materials Today Bio, 10*, 1–18.
29. Shadwick, R. (1999). Mechanical design in arteries. *Journal of Experimental Biology, 202*(23), 3305–3313.
30. Wolinsky, H., & Glagov, S. (1964). Structural basis for the static mechanical properties of the aortic media. *Circulation Research, 14*(5), 400–413.
31. Karpenko, M., Stosiak, M., Prentkovskis, O., & Skačkauskas, P. (2024). Energy–Saving in hydraulic drives in experimental approach and biomimetric similarity. In *Advances in hydraulic and pneumatic drives and control* (pp. 269–279). Springer Nature Switzerland.
32. Karpenko, M., & Nugaras, J. (2022). Vibration damping characteristics of the cork-based compo-site material in line to frequency analysis. *Journal of Theoretical and Applied Mechanics, 60*(4), 593–602.
33. European standard. EN 854:2015. (2015). Rubber hoses and hose assemblies. Textile reinforced hydraulic type. Specification.
34. European standard. EN 853 1SN:2015. (2015). Rubber hoses and hose assemblies. Wire braid reinforced hydraulic standard type. Specification.
35. European standard. EN 857 1SC:2015. (2015). Rubber hoses and hose assemblies. Wire braid reinforced hydraulic slimline type. Specification.
36. Karpenko, M. (2021). Investigation of energy efficiency of mobile machinery hydraulic drives. Ph.D. dissertation (164 p.). Vilniaus Gedimino technikos universitetas.

Energy-Saving Effects from Reducing Vibrations in Hydraulic System Elements—Valves and Pipelines

This chapter will specifically analyze the research results of the proposed energy-saving way in hydraulic drives based on a reducing vibration. Additionally, the effects of energy-saving in hydraulic systems by applying proposed results overview in the current chapter. The current book chapter held present how implementation of knowledge can help to design more substantiable construction of machine hydraulic systems with avoiding vibration problems.

6.1 Analysis of the Valve's Energy Efficiency Based on the Research

The existing analyses of the impact of external mechanical vibrations on the hydraulic valve indicate the conditions under which additional vibrations of the hydraulic distributor slider may be aroused. It was noted that passive vibration isolation methods can be used to limit vibrations of the valve body and its control element (e.g., divider slider) [1]. For the purposes of mathematical description, a mathematical model of the movement of the proportional distributor slider, whose body was affected by external mechanical vibrations, was presented. This model is written in a system of Eqs. (5.3) in Chap. 5. This chapter will analyze the impact of external mechanical vibrations on flow losses through a proportional distributor and the impact of reducing the slider vibration amplitude on changes in flow losses through a vibrating proportional distributor. The vibration of the distributor spool can be limited using the methods presented in Sects. 5.1 and 5.2. In order to carry out the analysis mentioned above, it was assumed that the proportional

© The Author(s), under exclusive license to Springer Nature Switzerland AG 2024   149
M. Stosiak and M. Karpenko, *Dynamics of Machines and Hydraulic Systems*, Synthesis
Lectures on Mechanical Engineering, https://doi.org/10.1007/978-3-031-55525-1_6

Fig. 6.1 Diagram of a hydraulic system in which a vibrating proportional distributor operates: 1—pump, 2—overflow valve, 3—proportional distributor subjected to external mechanical vibrations, 4—vibrating simulator table, 5a, 5b—pressure pulsation measurement points, 6—hydraulic motor

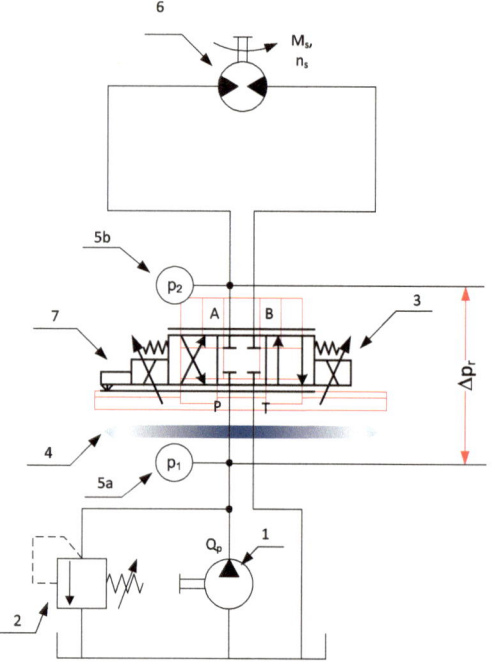

distributor stimulated to vibrate operates in the hydraulic system shown in Fig. 6.1. The diagram shows the pressure drop on the vibrating proportional distributor as Δp_r.

Numerical calculations were performed in the Matlab environment using the Simscape Fluids library. The symbolic diagram of the model implemented in Simscape Fluids is shown in Fig. 6.2. This environment requires parameterization of the hydraulic system [2]. During the calculations, the parameters of the hydraulic system were used: pump intake $Q_p = 6.5$ dm^3/min, average pressure behind the distributor 2 MPa, and the liquid used in the system is HL68 mineral oil. The vibration parameters of the simulator table with a proportional distributor placed on it are given under Table 5.1 in Chap. 5.

Pressure pulsation was subjected to numerical analysis for the frequency range of external mechanical vibrations acting on the proportional distributor from 20 to 60 Hz. The results are shown in Figs. 6.3, 6.4, 6.5, 6.6, 6.7, 6.8, 6.9, 6.10, 6.11 and 6.12. Figures 6.3, 6.5, 6.7, 6.9, and 6.11 show the forced oscillation motion for three values of the slider vibration amplitudes: a1 (blue) > a2 (orange) > a3 (gray). Amplitus a1 corresponded to the forced vibrations of the slider in the absence of vibration isolators—the value of the slider vibration amplitude corresponded to those obtained as a result of experimental tests presented in Chap. 4 (Fig. 4.15a). The amplitudes of forced vibrations of the a2 and a3 sliders were obtained by using vibration isolators to reduce vibrations of the distributor body and the slider.

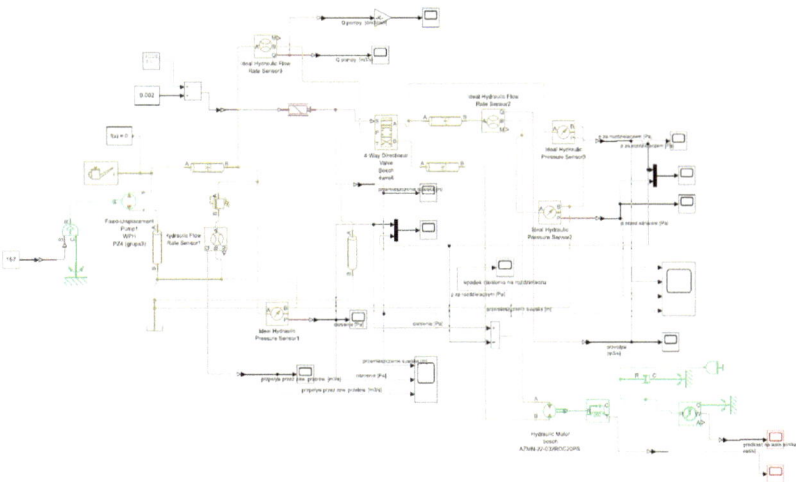

Fig. 6.2 Model of the hydraulic system in the center matlab/simscape fluids

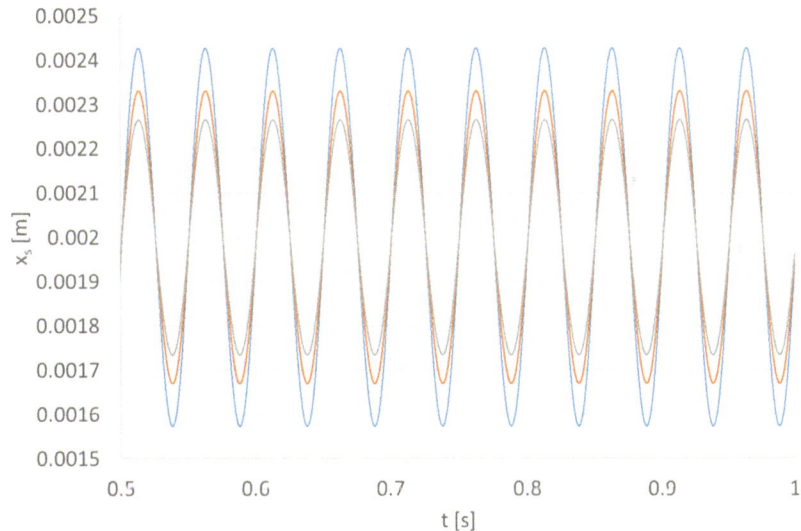

Fig. 6.3 Forced oscillation of the proportional distributor slider with a frequency of 20 Hz

In Fig. 6.4, the blue color of the pressure pulsation corresponds to the slider vibration amplitude marked in blue in Fig. 6.3(a1). The orange color of the pressure pulsation corresponds to the amplitude of the slider vibrations marked in orange in Fig. 6.3(a2). The pressure pulsation's gray color corresponds to the slider vibrations' amplitude, marked in gray in Fig. 6.3(a3).

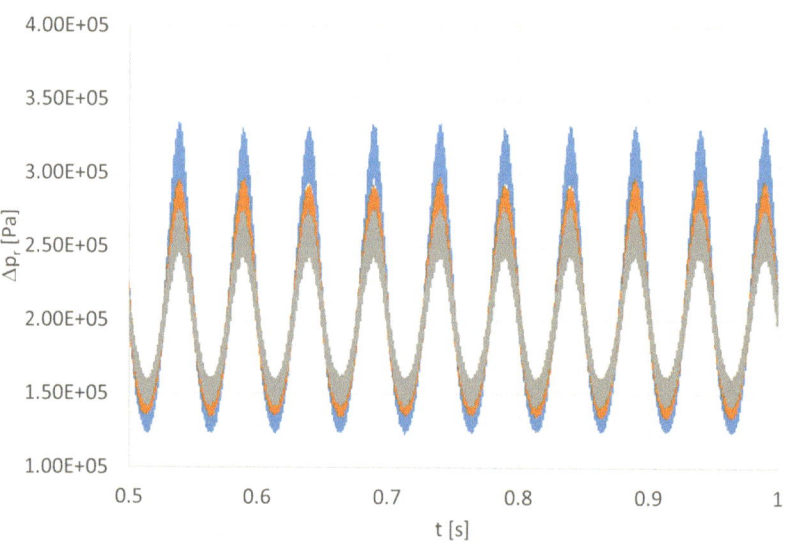

Fig. 6.4 Pressure drop Δp_r on the distributor when the slider vibrates with a frequency of 20 Hz

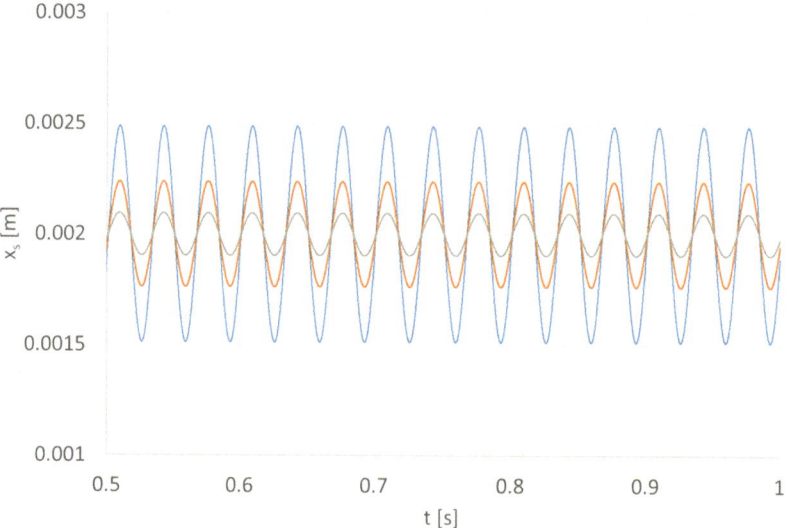

Fig. 6.5 Forced oscillation of the proportional distributor slider with a frequency of 30 Hz

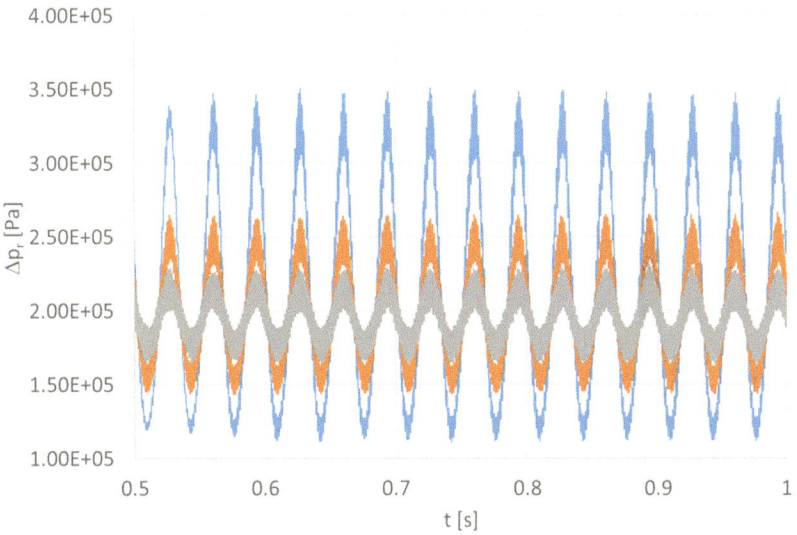

Fig. 6.6 Pressure drop Δp_r on the distributor when the slider oscillates with a frequency of 30 Hz

In Fig. 6.6, the blue color of the pressure pulsation corresponds to the slider vibration amplitude marked in blue in Fig. 6.5(a1). The orange color of the pressure pulsation corresponds to the amplitude of the slider vibrations marked in orange in Fig. 6.5(a2). The pressure pulsation's gray color corresponds to the slider vibrations' amplitude, marked in gray in Fig. 6.5(a3).

In Fig. 6.8, the blue color of the pressure pulsation corresponds to the slider vibration amplitude marked in blue in Fig. 6.7(a1). The orange color of the pressure pulsation corresponds to the amplitude of the slider vibrations marked in orange in Fig. 6.7(a2). The pressure pulsation's gray color corresponds to the slider vibrations' amplitude, marked in gray in Fig. 6.7(a3).

In Fig. 6.10, the blue color of the pressure pulsation corresponds to the slider vibration amplitude marked in blue in Fig. 6.9(a1). The orange color of the pressure pulsation corresponds to the amplitude of the slider vibrations marked in orange in Fig. 6.9(a2). The pressure pulsation's gray color corresponds to the slider vibrations' amplitude, marked in gray in Fig. 6.9(a3).

In Fig. 6.12, the blue color of the pressure pulsation corresponds to the slider vibration amplitude marked in blue in Fig. 6.11(a1). The orange color of the pressure pulsation corresponds to the amplitude of the slider vibrations marked in orange in Fig. 6.11(a2). The pressure pulsation's gray color corresponds to the slider vibrations' amplitude, marked in gray in Fig. 6.11(a3).

The analysis of the results presented in Figs. 6.3, 6.4, 6.5, 6.6, 6.7, 6.8, 6.9, 6.10, 6.11 and 6.12 shows that reducing slider vibrations reduces the pressure drop occurring

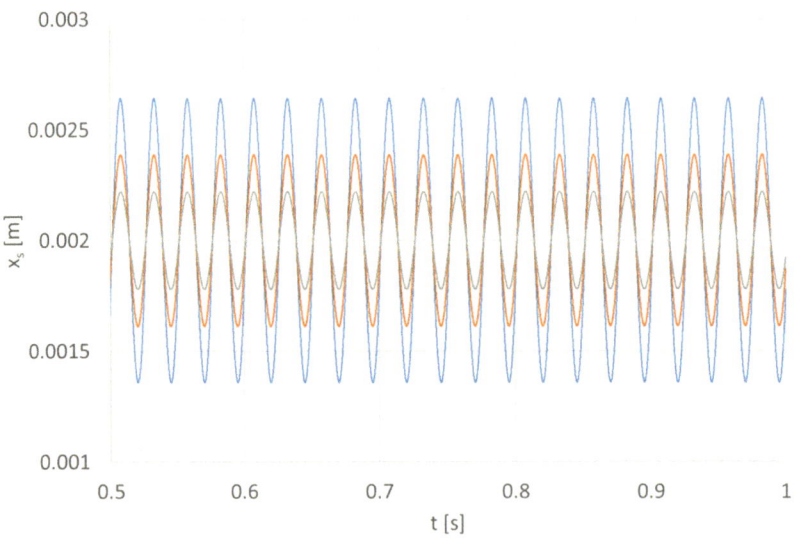

Fig. 6.7 Forced oscillation of the proportional distributor slider with a frequency of 40 Hz

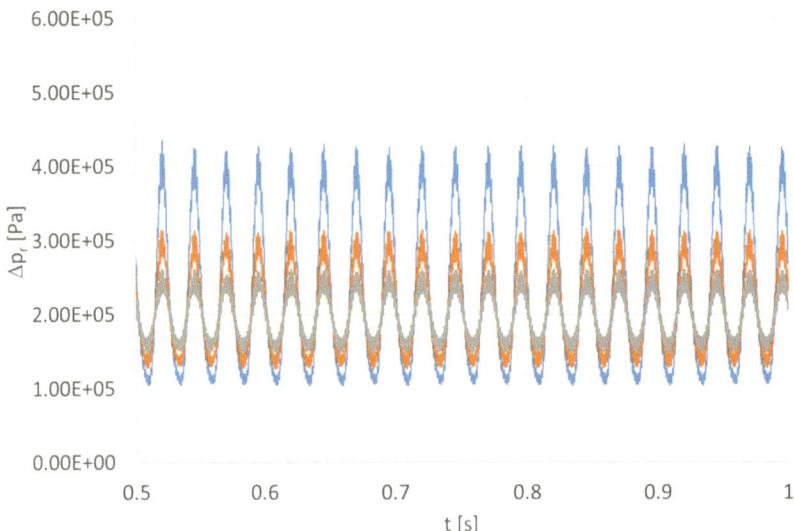

Fig. 6.8 Pressure drop Δp_r on the distributor when the slider vibrates with a frequency of 40 Hz

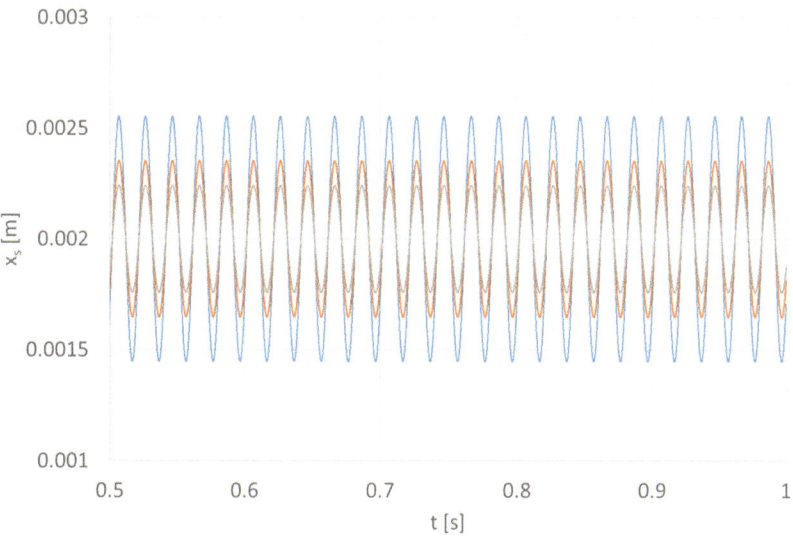

Fig. 6.9 Forced oscillation of the proportional distributor slider with a frequency of 50 Hz

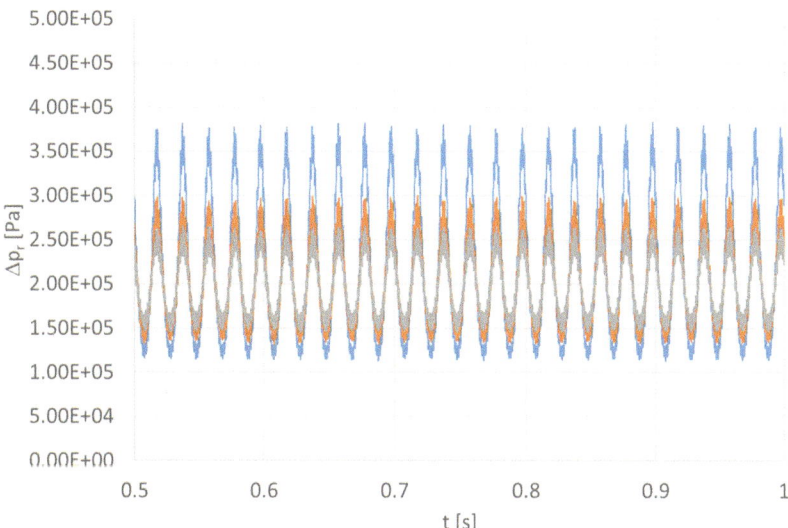

Fig. 6.10 Pressure drop Δp_r on the distributor when the slider vibrates with a frequency of 50 Hz

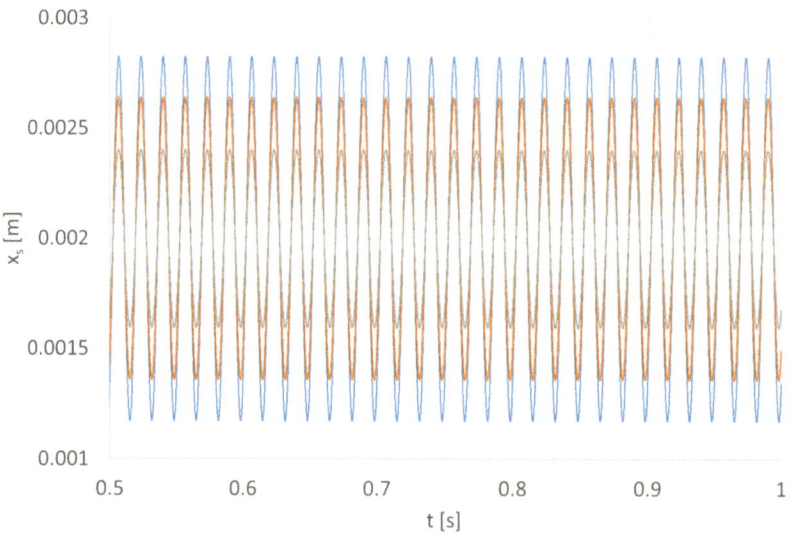

Fig. 6.11 Forced oscillation of the proportional distributor slider with a frequency of 60 Hz

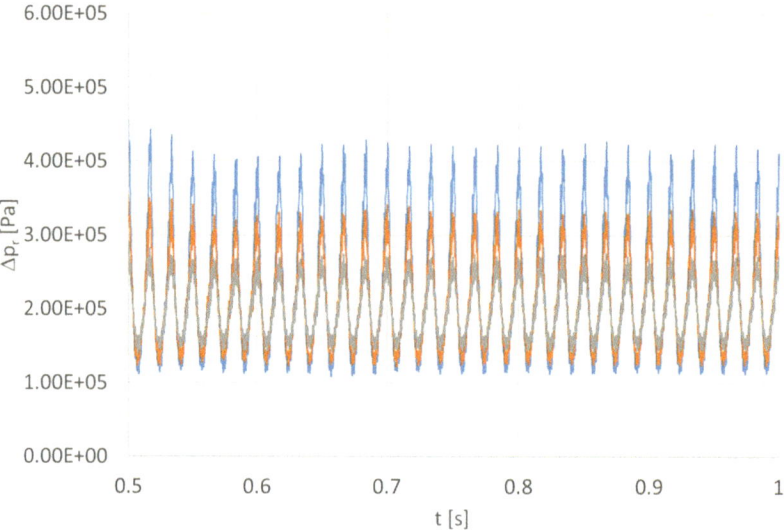

Fig. 6.12 Pressure drop Δp_r on the distributor when the slider vibrates with a frequency of 60 Hz

on the vibrating proportional distributor. Considering the liquid flow rate (its average value, constant over time), we can talk about reducing power losses on the vibrating distributor. The time graphs of power lost on the vibrating distributor will correspond to the graphs of pressure losses on the distributors, and the flow rate value will scale the values. Therefore, the greater the vibrations of the distributor slider, the greater the power losses in the distributor.

It is, therefore, advisable to limit the influence of external mechanical vibrations on the distributor body and its slider. These treatments lead to a reduction in the amplitude of slider vibrations and a reduction in pressure losses during the liquid flow through the distributor.

6.2 Analysis of the Pipeline's Energy Efficiency

An analysis of the energy efficiency of the pipeline is now demonstrated via a comparison of theoretical power losses for hydraulic drives, for which each pipeline is considered to have a one-meter length. The results are displayed in Fig. 6.13.

The efficiency deployed by using different pipelines in the hydraulic drive (symbolized as $\eta_{p-sp.}$) can be found using [3]:

$$\eta_{p-sp.} = \frac{\Delta N_{pipe}}{N_{drive}} \cdot 100\%, \tag{6.1}$$

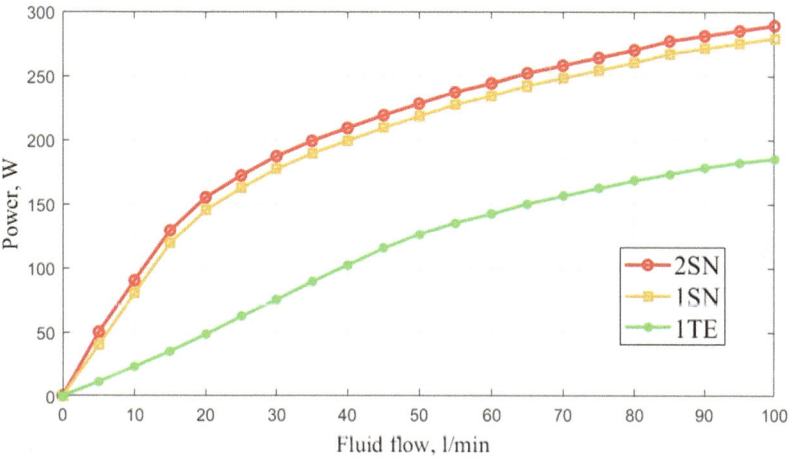

Fig. 6.13 Comparison of the power losses in the hydraulic drive for various pipelines with a one-meter length

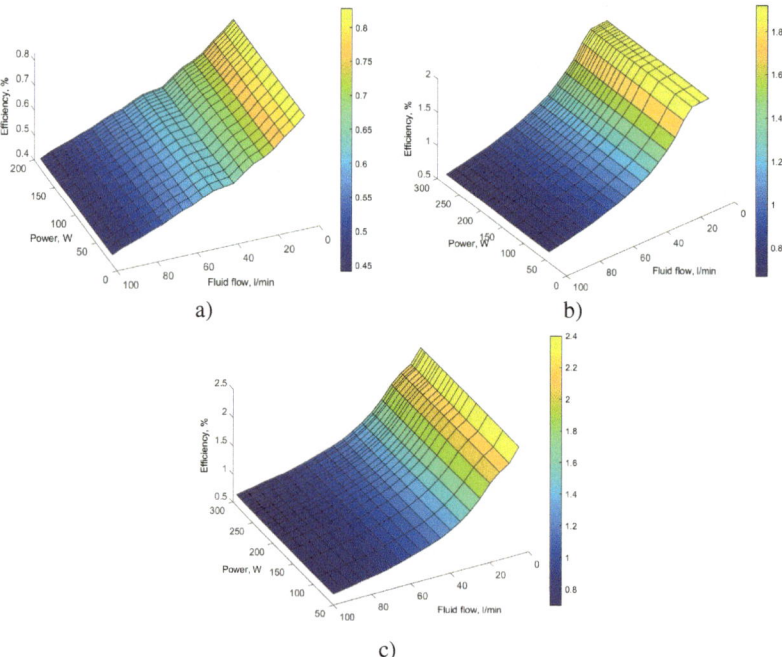

a)

b)

c)

Fig. 6.14 Graphs showing the efficiency deployed when using different pipelines in a hydraulic drive (with a pipeline of one-meter length): **a** 1*TE* pipeline type, **b** 1*SN* pipeline type, and **c** 2*SN* pipeline type

where ΔN_{pipe} denotes the power losses using different high-pressure hoses, and N_{drive} is the full power of hydraulic drive W.

Graphs showing the efficiency deployed when using different pipeline types in the hydraulic drive (with a pipeline of one-meter length) are presented in Fig. 6.14.

We now assess the energy efficiency of different pipelines based on previously documented measurement data. The calculation of the energy consumption E_c for a hydraulic pipeline can be determined using the following [4]:

$$E_c = \Delta P_i(t) \cdot Q_i(t), \qquad (6.2)$$

where $Q_i(t)$ is the fluid flow (in units of m^3/s) and $\Delta P_i(t)$ is the pressure losses (in units of Pa). In addition, we find:

$$\Delta P_i(t) = P_{inlet}(t) - P_{outlet}(t). \qquad (6.3)$$

where $P_{inlet}(t)$ and $P_{outlet}(t)$ are the fluid pressure at the inlet and outlet hose (in units of Pa).

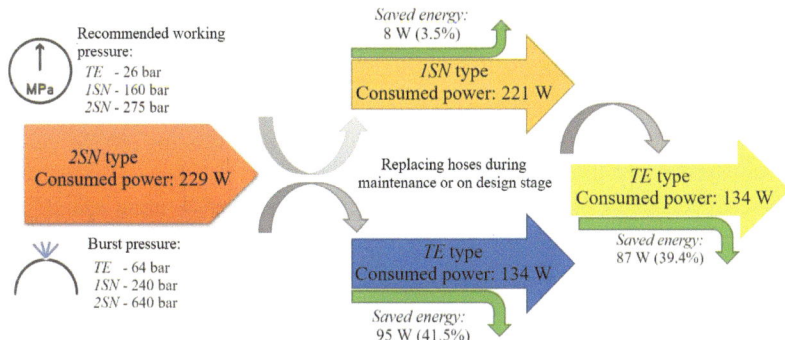

Fig. 6.15 Energy flow chart for pipeline consumption using the energy-saving analysis

Energy charts have been generated based on archived findings; these are presented in Fig. 6.15. These visual representations depict the transformation of energy (both visually and quantitatively) during the utilization or replacement of the high-pressure hoses in the hydraulic drives for mobile machinery. It is important to highlight that these outcomes were derived for a high-pressure hose with a length of one meter. The energy consumed in each hose was computed by multiplying the flow rate with the measured pressure difference at the hose's initiation and conclusion. For a detailed explanation of Fig. 6.15, the relationships (Eqs. 6.2 and 6.3) can be employed. The pressure loss values for each hose can be discerned from the pressure-time profiles depicted in a previous chapter. The energy flow charts are designed to visually and quantitatively depict the energy transformations during the replacement of the pipelines, specifically the high-pressure hoses in the hydraulic drives. Figure 6.15 illustrates the energy saved by substituting one type of pipeline with another.

Following the manufacturing recommendation, it is advised to employ 1TE pipeline, particularly up to a pressure of 26 bar. The obtained results indicate that optimal energy savings in the hydraulic drive under consideration can be achieved with the 1TE pipeline type. Compared to the 2SN and 1SN pipeline types, utilizing the 1TE pipeline type can result in savings of approximately 95 W and 87 W, respectively, per meter. This translates into a percentage savings of approximately 41.5% and 39.4%, respectively. If the hydraulic drive operates at pressures exceeding 26 bar, choosing the 1SN pipeline type (instead of the 2SN pipeline type) can yield savings of around 8 W or 3.5% per meter of pipeline length. The conducted research reveals that the 1TE pipeline type exhibits the highest efficiency (efficiency ranging from 0.8% to 0.45%), followed by the 1SN pipeline type (1.8–0.8%) and the 2SN pipeline type (2.4–0.8%). The investigation findings confirm that the *TE* pipeline type emerged as the most efficient choice for conserving energy in the examined hydraulic drive. In conclusion, the study suggests that substituting hydraulic pipes with alternatives that employ biomimetic approaches can diminish power losses and enhance the overall efficiency of hydraulic drives. Implementing such replacements could be seamlessly integrated into hydraulic drive maintenance routines or incorporated at the machinery design phase.

This chapter indicates that there are opportunities to influence the power losses in hydraulic systems by reducing the effect of vibrations on hydraulic valves and selecting the material structure of the hydraulic lines. The energy-saving search procedures should cover many aspects of the hydraulic system, which may lead to a cumulative effect in the reduction of power losses.

References

1. Piersol, A., & Paez, T. (2009). *Harris' shock and vibration handbook* (6th ed., 1168 p.). McGraw-Hill Handbooks.
2. Simscape Fluids. Model and simulate fluid systems—Documentation. https://www.mathworks.com/help/hydro/
3. Karpenko, M., Stosiak, M., Prentkovskis, O., & Skačkauskas, P. (2024). Energy–Saving in hydraulic drives in experimental approach and biomimetric similarity. In *Advances in Hydraulic and Pneumatic Drives and Control: International Scientific-Technical Conference on Hydraulic and Pneumatic Drives and Control NSHP 2023* (pp. 269–279). Springer. ISBN 9783031430015.
4. Karpenko, M. (2021). Investigation of energy efficiency of mobile machinery hydraulic drives (164p.). Dissertation, Vilnius Gediminas Technical University. https://doi.org/10.20334/2021-028-M

Summary and Notes on Criteria for Effective Vibration Reduction of Hydraulic Valves

7

The published research presented herein suggests that there is a correlation between external mechanical vibration acting on the hydraulic directional valve and the excitation of vibration in the directional valve spool, and consequently, as a periodically varying geometry of the flow gap in the spool pair, pressure pulsation. Additionally, this unfavorable phenomenon is significantly affected by the range of resonant frequencies of the spool, the frequency of external mechanical vibrations, and the angle between the direction of the spool's movement and the direction of the forcing vibrations—Chap. 4. Based on the presented results, it can also be stated that there is a correlation between the induction of pressure pulsations and external mechanical vibrations acting on a conventional single-stage directional valve and a single-stage relief valve—Chap. 2 and Appendix. This regularity can also be observed in hydraulic microvalves.

In order to reduce the adverse effects of external mechanical vibrations on hydraulic valves, it is important to indicate the individual transmission pathways of these vibrations to the valve control element and determine the dominant frequencies. The conducted research and analyses revealed that in typical lift valves (e.g., relief valves), the spring that presses the control element against the socket is responsible for the transfer of mechanical vibration from the valve body to the control element (e.g., ball, poppet). Herein, a spool pair of a typical hydraulic directional valve is theoretically analyzed and experimentally verified. A proposed modified description of the vibrating motion of the spool has been reported, considering elements of the mixed friction theory, as opposed to the previously commonly used Newtonian friction model. The proposed models of vibrating motion given by Eqs. (5.14), (5.23), (5.29), and (5.30) are parameterized based on the experimental tests performed.

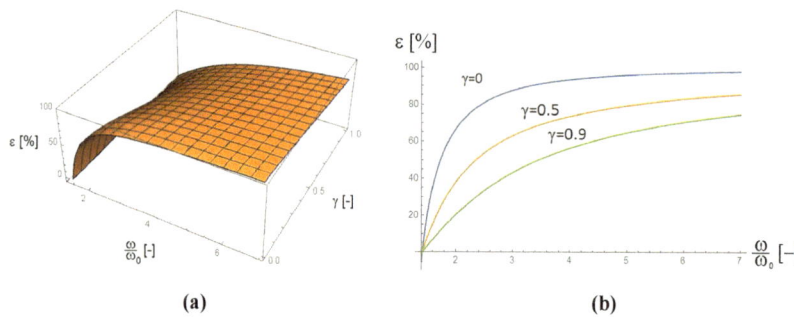

Fig. 7.1 The dependence of the efficiency of vibration isolation ε on the ratio between ω/ω_0: **a** dimensionless damping coefficient; **b** fixed value of the dimensionless damping coefficient

   The publication highlights methods of limiting the effects of external mechanical vibration on hydraulic valves by using passive vibration isolation methods that ensure high energy dissipation at appropriately high frequencies. The energy dissipation increases with decreasing stiffness of the vibration isolation. According to the previous considerations in Chap. 5, the following generalized conclusions are drawn as to the effectiveness of the vibration isolation system:

(a) If the condition $\omega \ll \omega_0$ holds for kinematic forcing, then the amplitude of the valve's steady vibration is close to the amplitude of the excitation vibrations, and for excitation by force, the amplitude of the force transmitted to the ground is comparable to the amplitude of the excitation force. Hence, vibration isolation is ineffective.
(b) When $\omega \approx \omega_0$, then the amplitude of the force transmitted to the ground increases in relation to the amplitude of the excitation force. Similarly, when a kinematically forced system is analyzed, the amplitude of the valve's absolute displacement increases in relation to the amplitude of the excitation displacement, indicating that the value of the amplification coefficient T_a (5.41) is > 1;
(c) Effective vibration insulation is considered when the ratio ω/ω_0 is greater than $\sqrt{2}$. This means that the value of the amplification factor is <1 for any attenuation condition. However, the effectiveness of the vibration isolation increases with decreasing damping—Fig. 7.1 and Table 7.1;
(d) When the excitation frequency further increases in relation to the frequency of natural vibration of the valve body, then the vibration isolation efficiency increases further and reaches almost 100%. However, the rate of increase diminishes when the ratio ω/ω_0 is greater than 5; for $\omega/\omega_0 \gg \sqrt{2}$ the impact of the change in damping on the effectiveness of vibration isolation decreases significantly, Table 7.2.
(e) The amplification factor can be reduced by increasing the damping in the range of resonance. If a machine or equipment fitted with hydraulic valves is expected to operate

Table 7.1 Vibration isolation efficiency ε for selected ratios of ω/ω_0 and dimensionless damping coefficient γ

ω/ω_0 [−]	ε [%] for $\gamma = 0$	ε [%] for $\gamma = 0.5$	ε [%] for $\gamma = 0.9$
2.5	81	54	34
3.5	91	69	50
4.5	95	77	61
5	96	79	65
5.5	97	81	68
6	97.1	83	70

Table 7.2 The efficiency of vibration isolation

The efficiency of vibration isolation ε%	The maximum value of transmission coefficient T	Required ratio ω/ω_0
90	0.1	3.32
80	0.2	2.45
70	0.3	2.08
60	0.4	1.87
50	0.5	1.73

for a long time in the resonance range of the valve body, a material with high damping should be chosen for the vibration isolator to reduce the valve body's vibration amplitudes. This is an excitation for the valve control element; in the resonance range, the value of the amplification factor T_a can be reduced by damping;

(f) If the dimensionless damping coefficient γ is greater than $\sqrt{2}/2$, then, in relation to relative steady-state vibrations described by the transmission coefficient B_a (5.29), the vibration is isolated effectively in the entire range of ratio ω/ω_0, i.e., for each ratio value, $B_a < 1$—Fig. 7.2. Note, in this case, the dimensionless damping coefficient is a function of, e.g., vibrating mass of an isolated element (e.g., directional spool), the stiffness of the spring pack (e.g., springs centering the spool) and stiffness of the vibration isolation element, damping in the spool-sleeve pair, and damping of the vibration isolation material. However, when the value of the dimensionless damping coefficient is $\gamma < \sqrt{2}/2$, then the effective vibration isolation occurs in the range of the ratio $\omega/\omega_0 \in \left(0, \dfrac{1}{\sqrt{2\cdot(1-2\cdot\gamma^2)}} \right)$, compared to steady-state vibration described by the transmission coefficient B_a (5.29);

Principles of vibration isolation are discussed using the example of a system with one degree of freedom. In practice, the system may have more degrees of freedom and may

Fig. 7.2 The waveform of the coefficient B_a as a function of the ratio ω/ω_0 for different values of the dimensionless damping coefficient γ

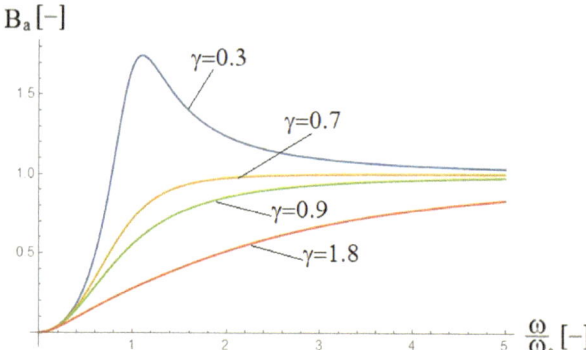

be affected by excitation forces with a wide spectrum of frequencies. In such cases, all-natural frequencies or those most dangerous in terms of resonance are identified (e.g., close to the resonance of the directional valve spool). By selecting the stiffness of the vibration isolation systems, natural frequencies can be shifted.

On the basis of theoretical analyses, a simplified and generalized scheme for the selection of isolating hydraulic valves from vibration is provided [1], displayed on Fig. 7.3.

Selection of the parameters of the vibration isolating system in the indication of the areas of vibration amplification can be carried out on the basis of the nomogram [1, 2] displayed on Fig. 7.4.

The efficiency of vibration isolation shown in Table 7.2 [1].

The guidelines and conclusions given are application-ready. The vibration isolation systems can also be designed by formulating a calculation algorithm using optimization methods and numerical computational techniques [3–5]. Among the many known optimization methods, the gradient method and the sensitivity analysis method are the basic ones [6–8]. In the process of vibration insulation system optimization, the static and dynamic features are selected in terms of reducing the amplitudes of displacements or accelerations of the hydraulic valve isolated from vibrations.

Fig. 7.3 Algorithm for selection of hydraulic valve vibration isolation

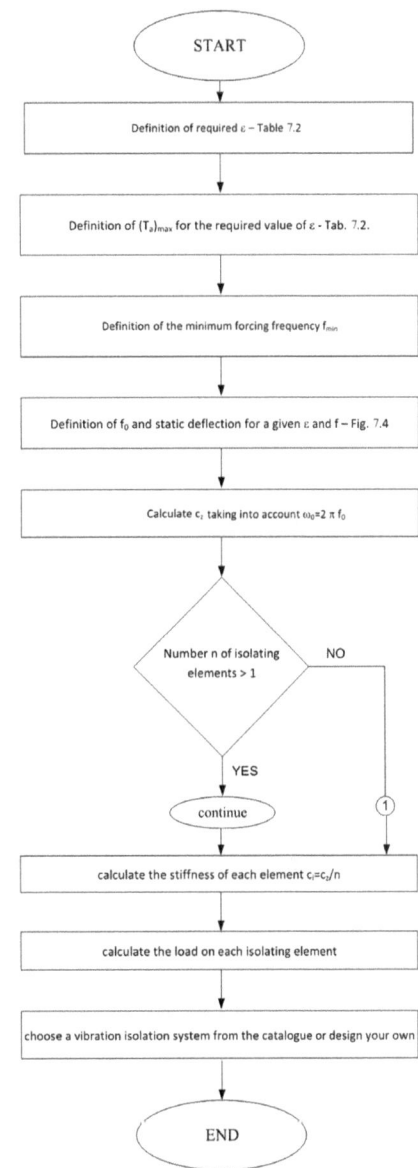

In gradient optimization, a defined integral function is considered the objective function [3]. In the case of the vibration isolation system of the hydraulic directional valve, the objective function can be, for example, the mean square criterion for the body vibration acceleration:

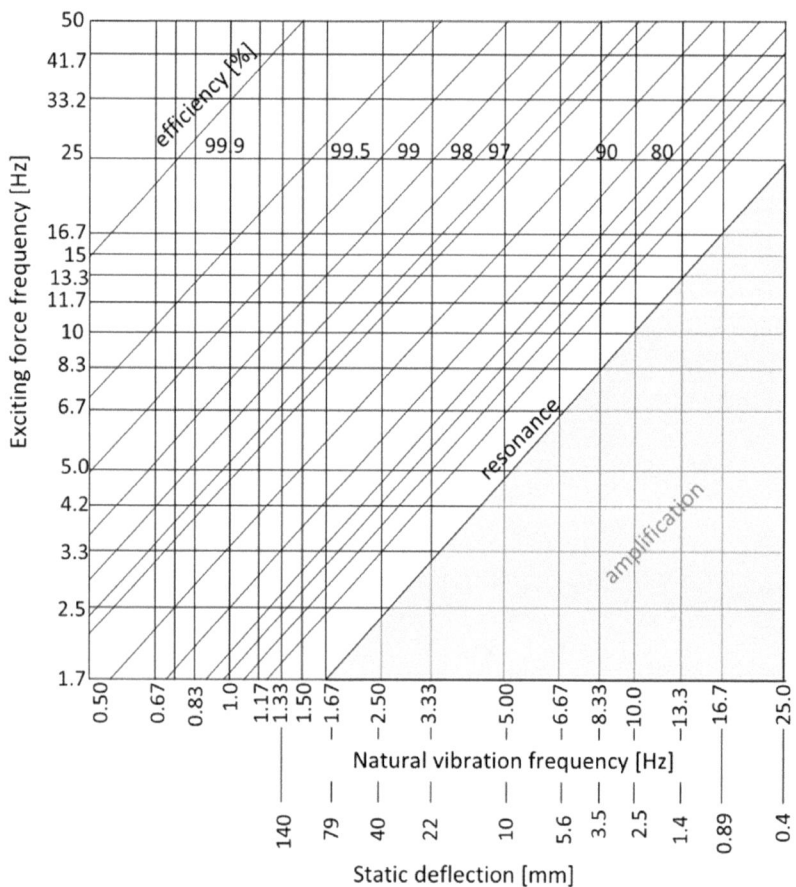

Fig. 7.4 Nomogram used in the process of selecting the parameters of the vibration iso-lation system

$$J = \sqrt{\frac{1}{T} \int_0^T |a_k(q, t)|^2 dt} \qquad (7.1)$$

where $a_k(q, t)$—acceleration of the valve body, q—set of design parameters of the vibra-tion isolator, e.g., $q = \{c_z, k_2\}$ for a vibration isolator with a linear stiffness and damping characteristics given as $c_z \cdot x + k_2 \cdot \dot{x}$ or $q = \{c_{1w}, c_{2w}, l, P_{2H}\}$ for a vibration isolator with quasi-zero stiffness. Assuming acceptable ranges of parameter variability:

$$c_{z\min} \leq c_z \leq c_{z\max}$$
$$k_{2\min} \leq k_2 \leq k_{2\max}$$
$$c_{1w\min} \leq c_{1w} \leq c_{1w\max} \tag{7.2}$$
$$l_{\min} \leq l \leq l_{\max}$$
$$P_{2H\min} \leq P_{2H} \leq P_{2H\max}$$

Their optimal values are sought, thereby, in the set of their values, the minimum of the objective function is determined using the gradient method, which has been described in detail by numerous publications [3, 4, 7].

When the sensitivity analysis method is employed to optimize the parameters of the vibration isolation system, the minimum of the function is identified iteratively in the parameter space, with small changes in their values, whereby the assumed values of acceleration of the distributor body are determined. As previously mentioned, the values of these parameters should meet the conditions (Eq. 7.2). Functional (Eq. 7.1) can also be adopted as the objective function. A defined logarithmic sensitivity function with respect to any parameter is tested. The algorithm for selecting the optimal parameters using this method is described by Ref. [3], and allows for the reduction of the functional value (Eq. 7.1) as well as achieving the assumed, permissible values of valve body accelerations.

A new approach for further reducing the impact of external mechanical vibration may be active vibration reduction systems. This involves the introduction of a structural or parametric modification to the vibrating system using an additional source of energy. However, such systems are more complex and require the introduction of automatic control. However, this is already possible for larger systems (such as a rotating machine, driver's seat, and buildings) that meet the often-mutually-exclusive requirements of vibration reduction efficiency in low-frequency ranges, a wide spectrum of forcing frequencies, stability, and dynamic stiffness. Due to their large dimensions, they are limited in terms of application, and cannot be used for the reduction of valve vibration. With regard to passive vibration isolation methods, materials should be sought that meet the requirements for elastic-dissipative properties while maintaining the requirement for oil resistance of the isolator material and the requirement to minimise dimensions.

Although the proposed methods of reducing the impact of external mechanical vibration on hydraulic valves can lead to the effective reduction of valve vibration or their controls, further reduction of the negative effects of vibrations can be achieved using pressure pulsation reduction methods [9]. Further reduction of the vibrational effects on hydraulic valves, or more specifically, the effects of vibration excitation of the control elements of these valves and the resulting pressure pulsation, is the appropriate selection of the length of hydraulic lines and considering the resonant phenomena. As indicated in Chap. 2, Sect. 2.4, it is possible, based on the quasi-steady friction loss model and the transmittances of the hydraulic long-line system defined by Eqs. (2.30)–(2.33) for pulsating flow, to reduce the amplitudes of pressure pulsations by selecting the length of the hydraulic lines.

Other methods of pressure pulsation reduction are based on structural modifications of the hydraulic system depending on the frequency range of pressure pulsations. The choice can be made from among a number of solutions of pressure pulsation dampers, e.g., broadband (chamber type, active) or selectively acting (e.g., branch type, chamber type, hydropneumatic accumulators) [10–14].

In relation to the effectiveness of the mechanical vibration reduction process, the best methods of reducing the effects of vibrations on hydraulic valves in the form of pressure pulsations are obtained by using several methods simultaneously.

In the area of the interaction of mechanical vibrations on valves and hydraulic systems, there remain many issues that require detailed research and description, followed by systematization. Some examples have been examined in this work, whereas others are still in their infancy and require significant development. Among the most important are:

- identification and pathways of vibration transmission from the vibrating valve through hydraulic lines to hydraulic receivers;
- development of new designs for hydraulic valves resistant to vibration;
- reduction of external forces by limiting the vibration of the machine or equipment fitted with hydraulic elements and systems;
- development (in the area of elastic-dissipative properties) of isolating materials that can be used in passive vibration isolation of valve bodies or their control elements;
- development of miniaturization, while maintaining the vibration isolation properties of active vibration reduction systems;
- the use of smart materials and materials with magneto or electro-rheological properties;
- investigations into the application of vibration-harvesting energy systems in relation to the vibrating elements of the hydraulic system.

The mutual, interrelated co-occurrence of mechanical vibration and pressure pulsations in hydraulic systems is the cause of a significant disadvantage of hydraulic systems, namely the noise of their operation. Therefore, it is expected that in future works, it is vital to examine those criteria listed above, as well as the reduction of the noise level of the hydraulic systems.

References

1. Engel, Z., Zawieska, M. (2010). Hałas i drgania w procesach pracy: źródła, ocean, zagrożenia. Centralny Instytut Ochrony Pracy – Państwowy Instytut Badawczy (in Polish).
2. Harris, C., Piersol, A. (2009). *Harris' shock and vibration handbook* (6th ed.). McGraw Hill. ISBN:10 0071508198.

3. Song, H., Shan, X., Hou, W., et al. (2023). A novel piezoelectric-based active-passive vibration isolator for low-frequency vibration system and experimental analysis of vibration isolation performance. *Energy, 278*, Part A, 127870. ISSN 0360-5442. https://doi.org/10.1016/j.energy.2023.127870

4. Tai, Y., Wang, H., Chen, Z. (2023). Vibration isolation performance and optimization design of a tuned inerter negative stiffness damper. *International Journal of Mechanical Sciences, 241*, 107948. ISSN 0020-7403. https://doi.org/10.1016/j.ijmecsci.2022.107948

5. Zhao, H., Feng, Y., Li, W., Xue, C. (2022). Numerical study and topology optimization of vibration isolation support structures. *International Journal of Mechanical Sciences, 228*, 107507. ISSN 0020-7403. https://doi.org/10.1016/j.ijmecsci.2022.107507

6. French, M. (2018). *Fundamentals of optimization.* Springer International Publishing AG. 331976191.

7. Law, M., Wabner, M., Colditz, A., et al. (2015). Active vibration isolation of machine tools using an electro-hydraulic actuator. *CIRP Journal of Manufacturing Science and Technology, 10*, 36–48. ISSN 1755-5817. https://doi.org/10.1016/j.cirpj.2015.05.00

8. Bakhshi, D., & Mehta, N. (2022). *Optimization techniques.* LAP Lambert Academic Publishing.

9. Kudźma, Z. (2012). Tłumienie pulsacji ciśnienia i hałasu w układach hydraulicznych w stanach przejściowych i ustalonych. Oficyna Wydawnicza PWr. (in Polish).

10. Kim, J., Yoon, G., Noh, J., et al. (2013). Development of optimal diaphragm-based pulsation damper structure for high-pressure GDI pump systems through design of experiments. *Mechatronics, 23*(3), 369–380. ISSN 0957-4158. https://doi.org/10.1016/j.mechatronics.2013.02.001

11. Kudźma, Z., & Stosiak, M. (2017). *Broadband pressure pulsation damper.* Patent nr PL 226524.

12. Yang, H., Liu, J., Yang, Z., et al. (2023). Active control of the fluid pulse based on the FxLMS. *Petroleum.* ISSN 2405-6561. https://doi.org/10.1016/j.petlm.2023.06.006

13. Bach, D., Masselter, T., & Speck, T. (2017). Damping of pressure pulsations in mobile hydraulic applications by the use of closed cell cellular rubbers integrated into a vane pump. *Journal of Bionic Engineering, 14*(4), 791–803. ISSN 1672-6529. https://doi.org/10.1016/S1672-6529(16)60444-4

14. Zuti, Z., Shuping, C., Huawei, W., et al. (2019). The approach on reducing the pressure pulsation and vibration of seawater piston pump through integrating a group of accumulators. *Ocean Engineering, 173*, 319–330. ISSN 0029-8018. https://doi.org/10.1016/j.oceaneng.2018.12.078

Appendix: Extended Results of the Experimental Analysis on the Impact of Mechanical Vibrations on Selected Hydraulic Valves

As part of the work, experimental studies were conducted on the effect of mechanical vibration on a single-stage relief valve DBDH 6 G18/100 from Mannesmann-Rexroth and an electrical-control single-stage 4/3 directional valve 4WE6E53/AG24N24 from Mannesmann-Rexroth. Selected research results are presented in Chaps. 2 and 4. Extended results of experiments are presented in this section. The diagram of the hydraulic system where the tested valves operated is shown in Fig. A.1.

The source of external vibrations that acted on the tested valve was the simulator of the linear hydrostatic drive Hydropax ZY25 which is described in Chap. 4. The excitation parameters (frequency and amplitude) were identical for the mentioned tested valves.

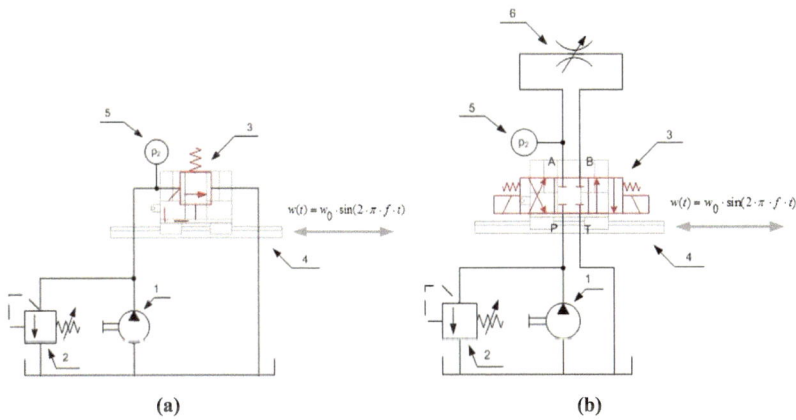

(a)   (b)

Fig. A.1 Diagram of the hydraulic system of the tested valve: 1—displacement pump, 2—safety valve, 4—vibrating table of the hydraulic simulator, 5—measuring point of the change in pressure using piezoelectric sensor M101A04 from Piezotronics: **a** 3—tested relief valve; **b** 3—tested directional valve 6—adjustable throttle valve

M. Stosiak and M. Karpenko, *Dynamics of Machines and Hydraulic Systems*, Synthesis Lectures on Mechanical Engineering, https://doi.org/10.1007/978-3-031-55525-1

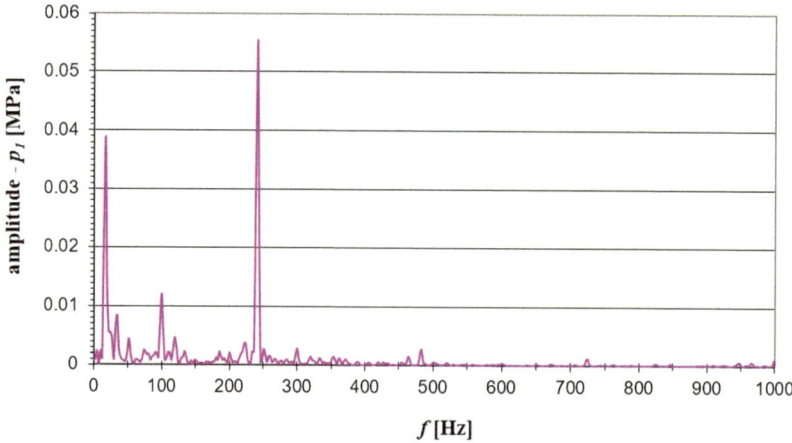

Fig. A.2 The amplitude-frequency spectrum of pressure pulsation in a hydraulic system with a relief valve excited by a frequency of $f = 15$ Hz

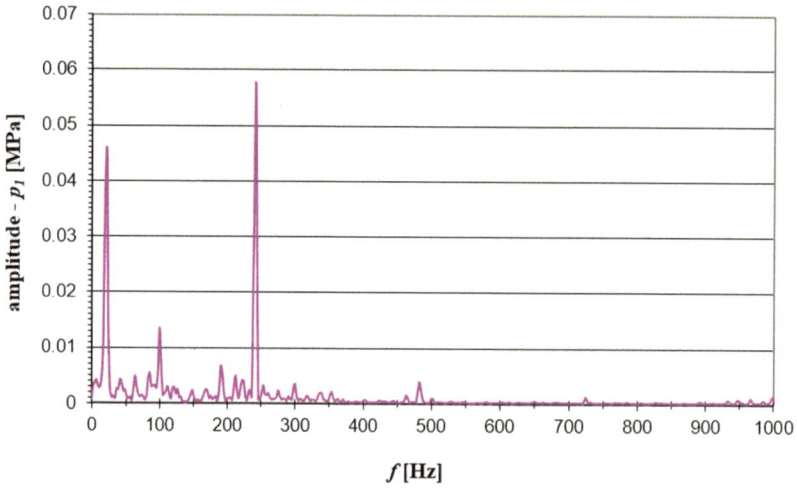

Fig. A.3 The amplitude-frequency spectrum of pressure pulsation in a hydraulic system with a relief valve excited by a frequency of $f = 20$ Hz

The diagrams in the further part of Appendix show the amplitude-frequency spectrum of pressure pulsation in a hydraulic system with a single-stage relief valve kinematically forced by a periodic function of the form $w = w_0 \cdot \sin(2\pi f t)$.

Analysis of the amplitude-frequency spectrum of pressure pulsations (in a hydraulic system with a kinematically excited relief valve) shown in Figs. A.2, A.3, A.4, A.5, A.6, A.7, A.8 and A.9 indicates the presence of a harmonic component with a frequency

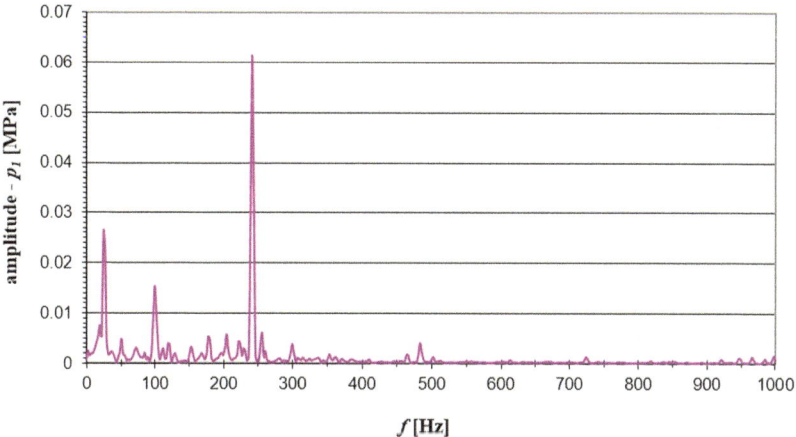

Fig. A.4 The amplitude-frequency spectrum of pressure pulsation in a hydraulic system with a relief valve excited by a frequency of $f = 25$ Hz

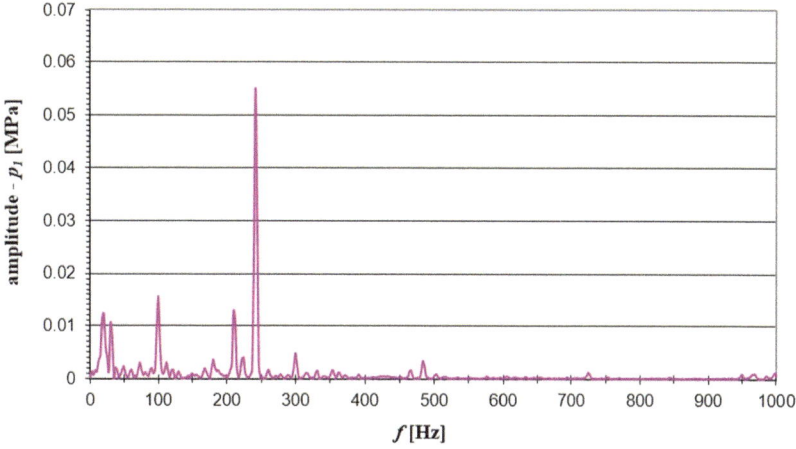

Fig. A.5 The amplitude-frequency spectrum of pressure pulsation in a hydraulic system with a relief valve excited by a frequency of $f = 30$ Hz

corresponding to the external mechanical vibration acting on the valve. This is particularly evident in Figs. A.2, A.3, A.4, A.5 and A.6, with the highest value of the component occurring for the excitation frequency of $f = 20$ Hz. This confirms the observations described in earlier chapters of this work (Chaps. 2 and 4). In the spectrum shown in Figs. A.2, A.3, A.4, A.5, A.6, A.7, A.8 and A.9, apart from the harmonic component corresponding to the frequency of the external excitation, there is a harmonic component resulting from the pulsation of the displacement pump capacity, which stems from the

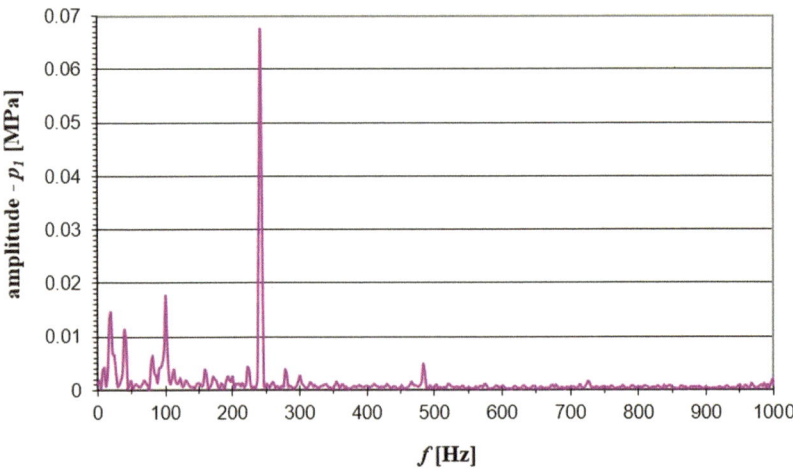

Fig. A.6 The amplitude-frequency spectrum of pressure pulsation in a hydraulic system with a relief valve excited by a frequency of $f = 40$ Hz

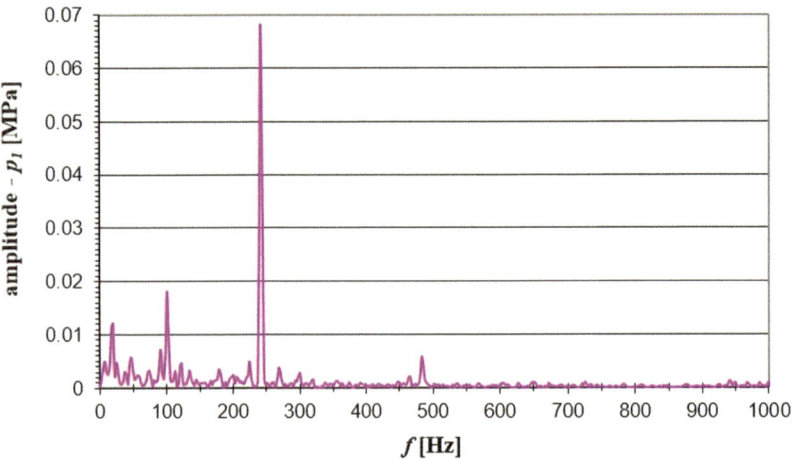

Fig. A.7 The amplitude-frequency spectrum of pressure pulsation in a hydraulic system with a relief valve excited by a frequency of $f = 45$ Hz

kinematics of the pump operation. The diagrams in a further part of Appendix show the amplitude-frequency spectrum of pressure pulsations in the hydraulic system with a 4/3 directional valve controlled by "on/off" coils, forced kinematically by a periodic function of the form $w = w_0 \cdot \sin(2\pi ft)$ (Figs. A.10, A.11, A.12, A.13, A.14, A.15, A.16 and A.17).

The research results on the impact of external mechanical vibration on pressure pulsation in a hydraulic system with a kinematically forced one-stage relief valve and a 4/3

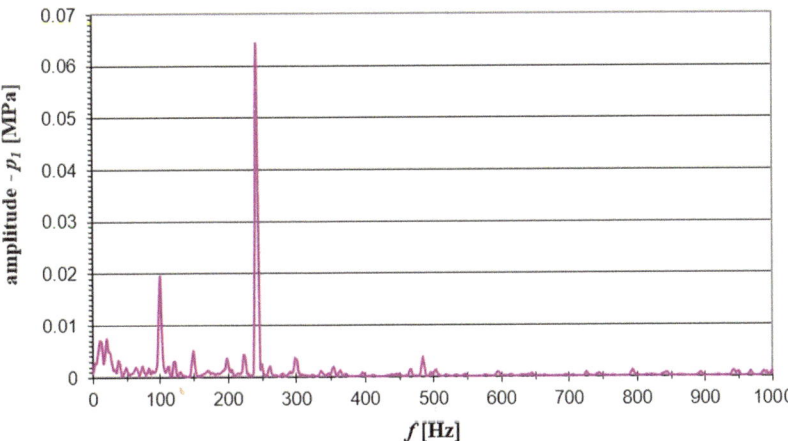

Fig. A.8 The amplitude-frequency spectrum of pressure pulsation in a hydraulic system with a relief valve excited by a frequency of $f = 50$ Hz

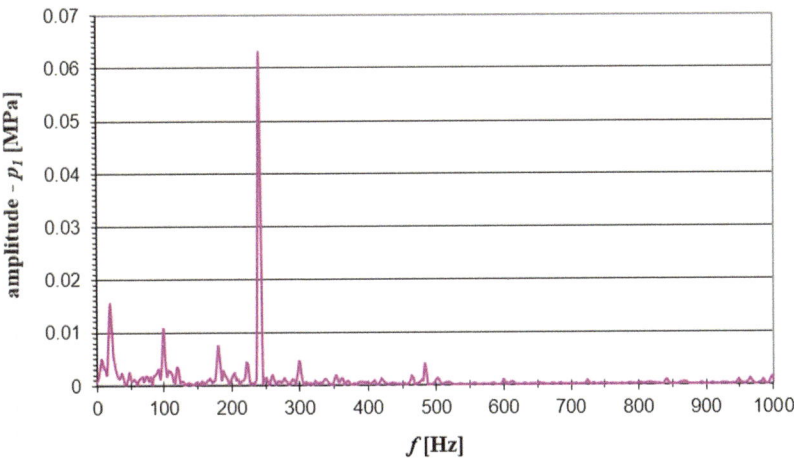

Fig. A.9 The amplitude-frequency spectrum of pressure pulsation in a hydraulic system with a relief valve excited by a frequency of $f = 60$ Hz

directional valve, included in Appendix 1, complement the research material contained in this work. This confirmed the adverse impact of mechanical vibrations on hydraulic components and systems, especially when the frequency of external mechanical vibrations is close to the resonant frequency of the valve control element (spool, poppet, etc.).

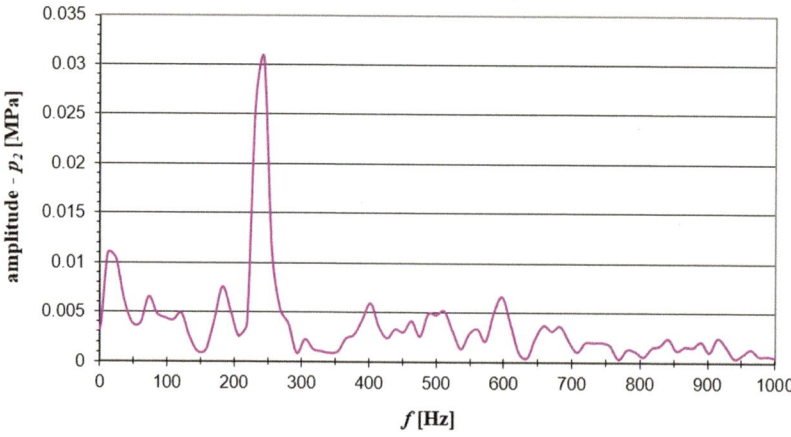

Fig. A.10 The amplitude-frequency spectrum of pressure pulsations in a hydraulic system with a directional valve forced by a frequency of $f = 15$ Hz

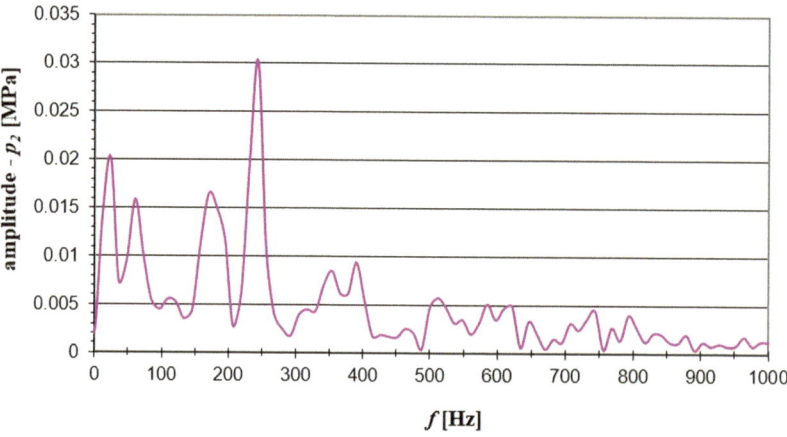

Fig. A.11 The amplitude-frequency spectrum of pressure pulsations in a hydraulic system with a directional valve forced by a frequency of $f = 20$ Hz

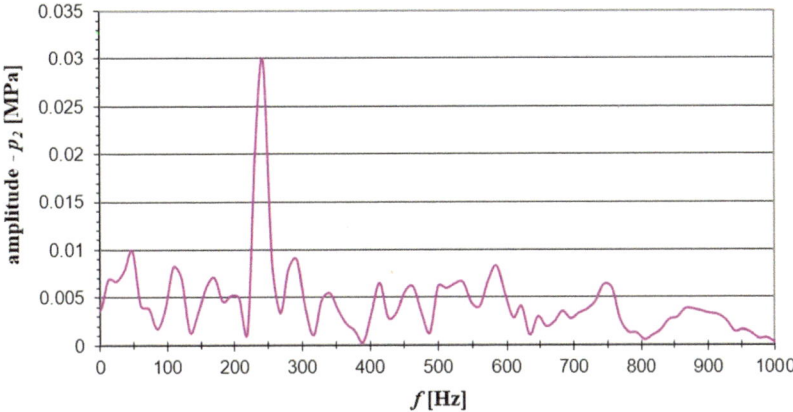

Fig. A.12 The amplitude-frequency spectrum of pressure pulsations in a hydraulic system with a directional valve forced by a frequency of $f = 25$ Hz

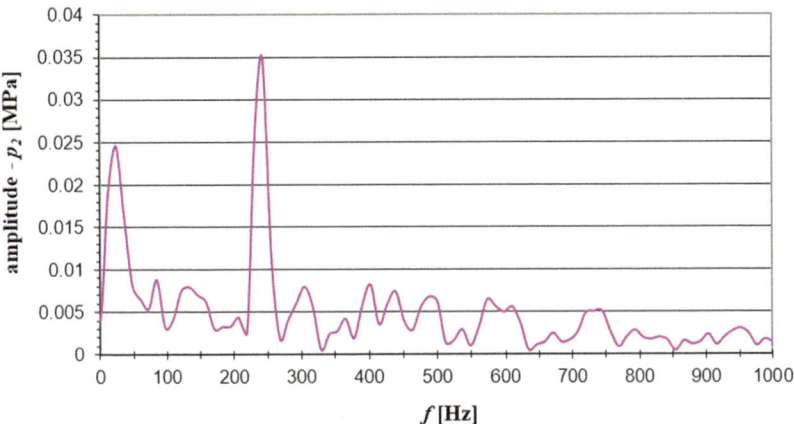

Fig. A.13 The amplitude-frequency spectrum of pressure pulsations in a hydraulic system with a directional valve forced by a frequency of $f = 30$ Hz

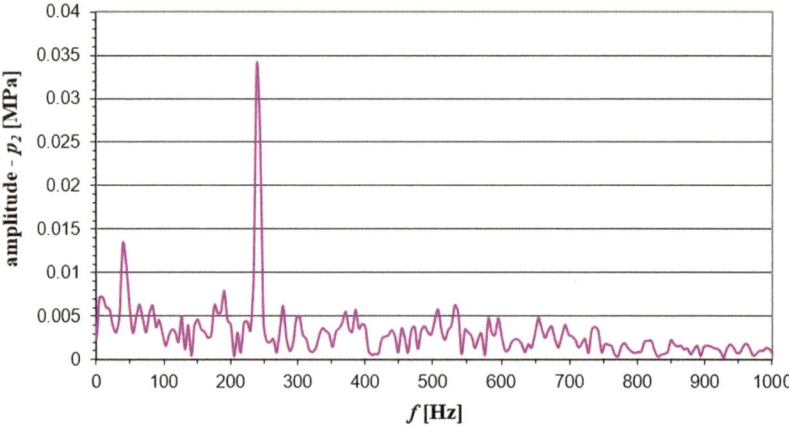

Fig. A.14 The amplitude-frequency spectrum of pressure pulsations in a hydraulic system with a directional valve forced by a frequency of $f = 40$ Hz

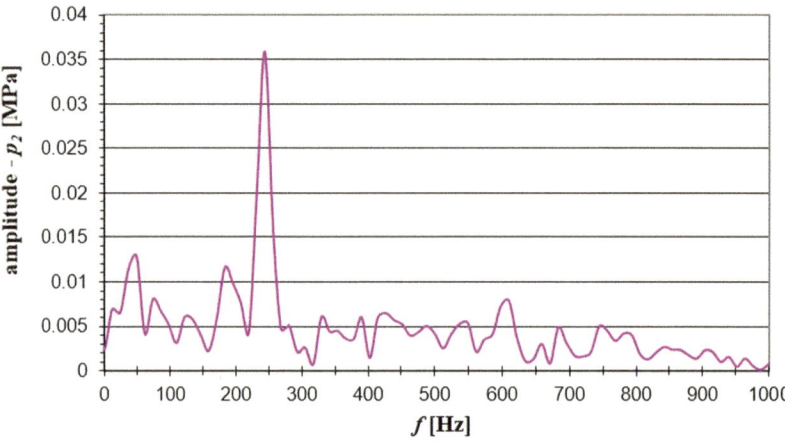

Fig. A.15 The amplitude-frequency spectrum of pressure pulsations in a hydraulic system with a directional valve forced by a frequency of $f = 45$ Hz

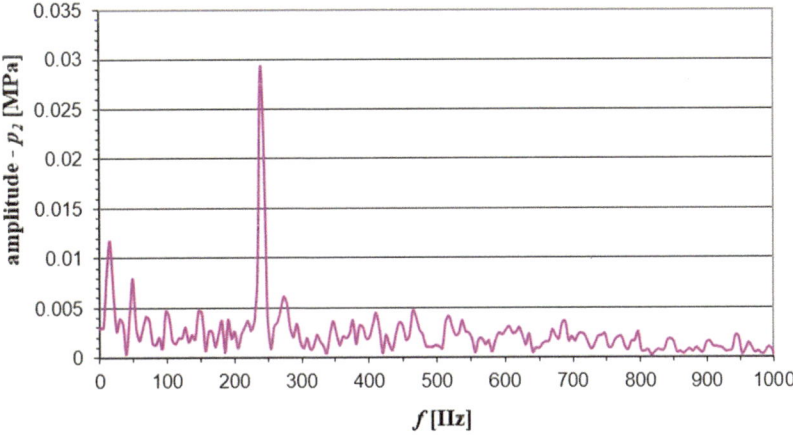

Fig. A.16 The amplitude-frequency spectrum of pressure pulsations in a hydraulic system with a directional valve forced by a frequency of $f = 50$ Hz

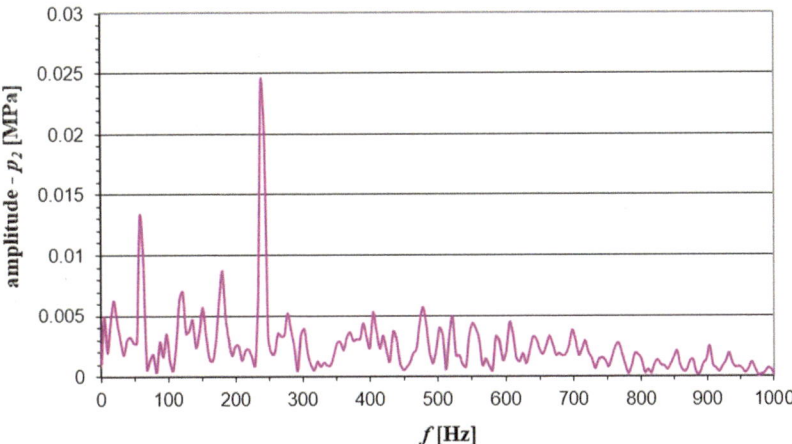

Fig. A.17 The amplitude-frequency spectrum of pressure pulsations in a hydraulic system with a directional valve forced by a frequency of $f = 60$ Hz